Praise for

Team Guide to Software Testability

"Finally, practical exercises with real-life examples and effective visual models so that your team can make real improvements to your product's testability. We've needed this book for decades. Get it today, dive into any chapter that seems to relate to your team's needs, and start improving one small step at a time."

— *Lisa Crispin, Quality Owner, OutSystems; co-author of 'Agile Testing'*

"Ash and Rob's vast experience shines through in this fantastic book. Packed with revealing activities, the book will give you a great set of tools to improve testability throughout your company. I love that the authors demonstrate the benefits to individual roles, providing the reader with clarity on why quality is a team responsibility."

— *Christopher Chant, Head of Production, Shelby Finance Limited*

"A must-read for anybody that wants to create quality software. The book covers all aspects of testability and gives readers everything they need to become testability-advocates themselves."

— *Joost van Wollingen, Head of Test Engineering, iptiQ*

"This book does an amazing job at describing testability. The practical exercises allow you to confidently explore testability concepts with your team. A must read for anyone involved in software development!"

— *Ali Hill, Consultant, ECS Digital*

Team Guide to Software Testability

Better software through greater testability

Ash Winter and Rob Meaney

Conflux Books

Leeds, UK

Team Guide to Software Testability

Ash Winter and Rob Meaney

Published by Conflux Books, a trading name of Conflux Digital Ltd, Leeds, UK.

Commissioning Editor: Matthew Skelton
Cover Design: Elementary Digital and Matthew Oglesby

For information about bulk discounts or booking the authors for an event, please email info@confluxbooks.com

ISBN	978-1-912058-66-2
eBook ISBN	978-1-912058-67-9
Kindle ISBN	978-1-912058-79-2
PDF ISBN	978-1-912058-69-3

Conflux Books: confluxbooks.com

Contents

Team Guides for Software

Pocket-sized insights for software teams

The *Team Guides for Software* series takes a *team-first approach* to software systems with the aim of **empowering whole teams** to build and operate software systems more effectively. The books are written and curated by experienced software practitioners and **emphasize the need for collaboration and learning, with the team at the centre**.

Titles in the *Team Guides for Software* series include:

1. *Software Operability* by Matthew Skelton, Alex Moore, and Rob Thatcher
2. *Metrics for Business Decisions* by Mattia Battiston and Chris Young
3. *Software Testability* by Ash Winter and Rob Meaney
4. *Software Releasability* by Manuel Pais and Chris O'Dell

 Find out more about the *Team Guides for Software* series by visiting: http://teamguidesforsoftware.com/

Conflux Books

Books for technologists by technologists

Our books help to accelerate and deepen your learning in the field of software systems. We focus on subjects that don't go out of date: fundamental software principles & practices, team interactions, and technology-independent skills.

Current and planned titles in the *Conflux Books* series include:

1. *Build Quality In* edited by Steve Smith and Matthew Skelton (B01)
2. *Better Whiteboard Sketches* by Matthew Skelton (B02)
3. *Internal Tech Conferences* by Victoria Morgan-Smith and Matthew Skelton (B03)
4. *Technical Writing for Blogs and Articles* by Matthew Skelton (B04)

 Find out more about the *Conflux Books* series by visiting: confluxbooks.com

Acknowledgements

We've learned and taken huge inspiration from many experts and practitioners whilst writing this book. We would like to thank the following people for their help and inspiration (in alphabetical order): Andy Butcher, Angie Jones, Christopher Chant, James Bach, Janet Gregory, Jon Allspaw, Lisa Crispin, Maria Kedemo, Mark Winteringham, Nicola Lindgren, Nicole Forsgren, Richard Bradshaw, Sally Goble, Steve Smith, Tom Hudson.

We also give a big "thank you" to our reviewers: Ali Hill, Christopher Chant, Dave Harrison, Joost van Wollingen, Lisa Crispin, Urbano Freitas.

Ash would also like to thank: Rob for being a constant font of ideas and models as well as a good friend; Gwen for reminding me that writing clearly is more important than the vocabulary; Manuel for guarding the former when it slipped past Gwen; Matthew for his drive to get the books moving and his ability to impart his growing knowledge of the publishing industry. Finally, a special thank you to Martin Gijsen for asking the right questions to focus my mind on what testability really means to me.

Rob writes: "I'd like to thank the testing community for supporting and encouraging me throughout the years, Matthew for the opportunity to share my ideas and Manuel for the guidance in expressing them. I'd like to thank Ash firstly for all his help is putting this book together but more importantly for being such a patient, wonderful person throughout the process. Lastly, I'd like to thank Tom and Mary for planting the seed and Miriam, Adam and Abigail for being my light".

– Ash Winter and Rob Meaney

Exercises

Introduction

Why is testability important?

How can software teams adapt and change the software systems they build and run? How can we get 'early warning' of performance, compatibility and integration problems? How can we design software architectures that enable ongoing evolution? The answer is to have a sustainable focus on testability.

Testability is a vital property of modern software. It enables software teams to make changes rapidly and safely. With clear feedback loops to understand the impact of changes. The testability of your software product links to its operability and maintainability. There is a close relationship between the extent to which the software meets customer needs and how testable it is. If you want to drive improvements in both speed and agility, testability is the fuel for organizations delivering modern software.

Team dynamics

Imagine if there was a constant focus on how to observe, control and understand the product within your team. Teams need all kinds of skills, and everyone brings different skills to the table. Developers need it to implement change confidently, testers need it to assess risk and provide actionable information, customers need it to achieve their goals with the product. Creating a focus on testability can help bring teams and their customers closer together.

Predictability

Have you ever been asked if your team could go a little bit faster? Or worse, as an individual on behalf of the team? We know we have. We believe that what your customers are *really* asking for is a little more predictability, delivering software consistently to a known level of quality. Being able to observe, control and understand a system describes predictability pretty well. If your system struggles with those three characteristics, how can you imagine delivering in a predictable fashion? Testability isn't the only piece of the predictability puzzle, but it's a big part.

Supporting your business and customers

A focus on testability helps to support many aspects of your business. Operations can control the system, product management can experiment with features for feedback, marketing and design can get feedback on who is using the product and if customer's needs are being met. Most products have peak consumption periods when many stakeholders use the system simultaneously. This could be due to a major music or sporting event, for example.

 A testable system provides information about its own limits, giving your business and technical stakeholders the ability to make crucial decisions at peak times, when reputations can be won or lost.

Fast feedback

For us, one of the real questions in software development is: how can we get meaningful feedback on what we have created? Be it quantitative feedback through usage statistics and financial measures, or qualitative interpretation of customer experience, enhancing testability can enrich this feedback.

This is also true for feedback on system failure. Meaningful information on failure modes enhances your ability to test a system and its time to recovery after an outage or error, which is crucial for those who support the system. When things go wrong, being able to observe and control the system will be invaluable to both business and technical operations.

Data, not opinion

Everyone in the team has opinions. However, focusing on testability as a team will help you go beyond opinions and have data to support you. Being able to objectively test the assumptions and hypotheses behind the functional and operational features you build helps establish a dialogue with product stakeholders. Enhancing testability allows quantifying the business and operational impact, while also moving conversations away from those who shout loudest to those with the right data.

What does hard-to-test feel like?

Take your mind back to 'that' product. The one which caused you quite a lot of pain, phone buzzing in the middle of the night, and yet another failed cron job. The product that caused a lot of distress to the poor operations people when you handed them a complete mess.

Do you recognize any of these traits?

- Interacting with a product gives you no feedback. No logs, no audit trail, only mysterious unmonitored dependencies. You don't know if anything went well. Or badly.

- Interacting with a product gives you vast amounts of feedback. Log files spam uncontrollably, lights twinkle on dashboards, integrated dependencies give strange, opaque answers. You don't know if anything went well. Or badly.
- You release your product. Scores of utterly baffling issues pop up. Seemingly unrelated but somehow intertwined. Next release makes you twitchy.

What does testable feel like?

We would like to evoke a different feeling when you talk about your next project or product. Imagine a product where:

- You are in control of the amount of feedback your product gives you, allowing either a deep, detailed view or a wide, shallow view. Rather than trying to parse what your product is doing, it will tell its story.
- The product can be set into a state of your choosing, whether that be data or configuration, allowing you to develop your product with much more certainty.
- After release, you are not dreading the call from support or account management that your customers are unhappy. Any problems are flagged early and can be proactively managed.

What leads to testability being neglected?

During our careers, we have seen various maturity models of the "'ilities" of software deployed within organizations. Testability might make an appearance every now and then but it rarely seems to rank as high as scalability or reliability. We hope to give strategies for overcoming apathy and some common constraints to testability:

- No paradigm - if you don't understand what it is, how can you ask for it? Never mind describing its benefits to those who are paying for the product development.
- Nobody knows who should be responsible for it - product people think it has nothing to do with them, developers think it's testers' responsibility, testers don't communicate the pain of a hard-to-test system to developers.
- There's no time - "we need to build the thing now", when the pressure is on, with deadlines looming, there is no time for testability.
- It's not a functional requirement - testability and other operational features never make it into the backlog. But these features are what turns functionality into a product.

What is covered in this book?

Testability is a vast topic, which permeates much of software development. Our focus will be on how the team can enhance their systems and interactions with stakeholders through a focus on testability. We will share our experiences, those of others we have encountered along the way and many practical resources and references we can use on our testability journey.

What we *will* cover includes:

- Improving interactions within and between teams to enhance testability both through conversation and documentation
- Practical approaches for creating testable architectures, addressing common risks and useful solutions
- Taming your environments from local, disposable environments all the way to live ones
- Keeping the focus on testability through the concept to customer and beyond the lifecycle of a product

Although this book touches on some technical aspects of software testing, it doesn't focus on how to make specific artifacts like legacy code or user interfaces more testable. There is a lot of great guidance out there for those, for example "Working Effectively with Legacy Code" by Michael Feathers (2004). Testability is not only about testers and, by extension, not solely about testing!

How to use this book

Ideally, each chapter is meaningful independently. You should be able to identify your situation and pick the thinking, tools and techniques that help you most in your current situation. However, like most books, this also represents our story, what we have learned over the years. Hopefully those lessons can help you with your lessons, as a holistic experience.

Concepts and theory are important, but we value practical examples highly in this book and have sourced them from various friends within the world of software development, as well as ourselves. These examples are critical to the book and will hopefully provide value to our readers.

Most of all, we believe the process of developing systems and products is a joyful, collaborative endeavor. We hope that shines through the book!

Why we wrote this book

As experienced testers, we have seen teams struggling with hard-to-test systems for a long time. This has usually manifested itself in pain for testers of all kinds so we felt it. In our experience, there has been little will do anything about it. Through lack of knowledge, advocacy and an acceptance that hard to test is normal. Team testing cultures are hard to create. Enhancing testability makes this journey much easier.

Looking around at the wider body of knowledge about testability, we wanted to fill some gaps. In particular, creating architectures designed for testability has light coverage. Once locked into a hard to test architecture, retrofitting testability is a challenging journey. We hope the book helps those working on systems of all ages to improve their architectures.

We are both testers who have looked beyond our role into other disciplines, engaged in the whole system, including people, processes, products and much more. Improving testability involves that wider system over optimizing for local gains. It's time for a new practical focus on testability and this book contains the tools for teams to do that.

Feedback and suggestions

We'd welcome feedback and suggestions for changes: please contact us at info@confluxbooks.com, via @TestabilityBook on Twitter, or on the Leanpub discussion:
https://leanpub.com/SoftwareTestability/feedback

Ash Winter & Rob Meaney - August 2021

1. Set a pragmatic direction for improving testability using trade-off sliders

---

Key points

- Testability is often neglected as a **first class concern** we must be able to answer the **challenges and questions** from a wide range of roles.
- Shifting towards greater testability means first **establishing a starting point** showing that testability is rooted in **everything a team does**.
- Attempting to improve testability without knowledge of what the **priorities are and where value lies** will bring you into conflict with stakeholders.

This chapter covers techniques for improving the testability of software by acknowledging and highlighting the trade-offs being made or being planned.

1.1 Overcome common challenges to setting a team testability focus

Let's be honest. There is often a reason not to focus on testability. Commonly, there is a new feature to deliver for a particular date. Which the team needs to build now and worry about how observable, controllable and transparently documented it is later. The problem is that later will regularly become never. See if you can recognize any of these:

Challenge	Response
Testability? That sounds like something that only testers should care about. As a developer, why should I care?	Focusing on testability enhances the feedback you can get on the code you produce. That might be information about how much monetary value it provides or how it performs in production
We are really keen to get this feature out before the marketing campaign. What does it matter how testable it is?	When deadlines are tight, keeping a testability focus is important. We want to be able to make the right call on when to release. If the feature needs lots of setup time to test, we will spend less time performing testing
We will think about testability later, when we have built something sufficient to begin end to end testing.	It's important to test the smallest thing possible to get feedback. Our code contains assumptions and bias which we should test as early as possible.

Challenge	Response
We should focus on the performance and scalability of the system, testability is not as important as those.	Testability is part of those factors and makes the testing in those areas more valuable. When a system is easier to observe and control, you can better find its operational limits. Without functional or state problems hindering you.
What I'd like to know is: how will testability make us more money and protect our reputation with our clients?	A system that is easier to test means reliable delivery of the features you want. At a more transparent level of quality. Plus, when you have an outage, testability can decrease time to recovery through better observability and system control mechanisms.
You can't test this change. It's a refactoring, library update or config change exercise which shouldn't have a functional impact.	If we feel we can't test the change, we don't understand the scope of the change and its potential impacts. Testability helps to map risk and impacts on our architecture, which helps us to better determine what to test.

Over our careers, we dealt with a lot of these questions and challenges, to varying levels of success. Being able to field them confidently helps when you start your journey to making testability a first-class concern.

1.2 Exercise: do the Team Test for Testability for a quick testability health check

We are big advocates of getting the right people in the room and getting them to collaborate. This generates buy-in from your stakeholders. This needs to be light touch, designed to get the team thinking about testability and where they are right now. Also note that none of the questions explicitly mention testability. This keeps the focus on what you do, rather than overloading with new concepts.

The Joel Test is a great example of this explicit approach (Spolsky2000). The Joel Test is a set of questions with yes or no answers that give you a picture of your current health as a team. As a first step into a team's testability journey, we have designed a similar test. It's called "The Team Test for Testability".

What does the test involve?

The test itself has 12 questions with binary answers. This is surprisingly difficult. Binary questions are hard but we are not trying to ignore context altogether. That will come later:

1. *If you ask the team to change their codebase do they react positively?*
2. *Does each member of the team have access to the system source control?*
3. *Does the team know which parts of the codebase are subject to the most change?*
4. *Does the team collaborate regularly with teams that maintain their dependencies?*
5. *Does the team have regular contact with the users of the system?*

6. *Can you quickly, easily and safely reproduce issues encountered by customers?*
7. *Is each member of the team able to create a disposable test environment?*
8. *Is each member of the team able to run automated unit tests?*
9. *Can the team test both the scheduled and on-demand parts of their system?*
10. *Does each member of the team have a method of consuming the application logs from production?*
11. *Does the team know what value the 95th percentile of response times is for their system?*
12. *Does the team curate a living knowledge base about the system it maintains?*

The beauty of the request for a yes or no answer is the ability to determine where team members differ. Plus, what they believe justifies their answer. One person's living knowledge base is a wiki page. For another it is an executable specification of acceptance tests. The perception of how testable a system is can be as important as how testable it actually is.

We believe the best testability strategies are balanced. These twelve questions touch on both internal and external factors which impact testability.

How does the test work?

In the example below, we have set out the test as an exercise for the team to come together and frame the testability of their system. This could be done as a survey if you prefer. For us, though, bringing the team together to talk about testability is a boon to testability in itself, so we advocate treating it as a group exercise.

Follow this facilitator's checklist below to help you.

Who do we need?

As this is the first step to building a wider culture of testability within the team, try to get contributors from the following groups:

- architects
    - enterprise architects
    - solution architects
- ops
    - system administrators
    - site reliability engineers
- testers
    - explorers
    - automators
- developers
    - back end
    - front end
    - full stack
- analysts
    - business
    - data
- product
    - owners
    - managers
    - users
- designers
    - UX
    - UI
- other development teams
    - internal
    - external

Preparation

- Time - One hour
- Type of Space - Not a boardroom, open space is preferable
- Attendee Preparation - Read the notes around each of the questions beforehand (Winter2017)
- Physical Materials
  - Copies of the 12 questions
  - A whiteboard or flipchart
  - Post-its and pens (one color)
- Digital Materials - None

Sample agenda for pre-communication

- Set the Stage (5 mins)
  - Goals, Outputs, FAQ
- Gather Data (10 mins)
  - Answer the 12 questions individually and anonymously
- Generate Insights (30 mins)
  - Map the responses
  - Discuss the differences and similarities

Facilitation

- Change the wording of the questions for common terminology at your organization if you need to
- The gather data step is better done anonymously, we want to learn about individual perceptions and the differences/similarities between them

Frequently Asked Questions

- What if someone wants to add a question?
  - Check if it's covered already, but mostly say yes!
- What if the attendee doesn't know the answer to a question?
  - Tell the attendee Mark as No. If you don't know about it, then it's the same as a no. All part of the learning

Goals

- Recognize the extent to which the team is putting focus on testability
- Understand the team's current perceptions of the testability of their own application

Outputs

- A bunch of question sheets with 12 yes or no answers on each
- A chart of how many yes or no answers there are
- A list of the key similarities and differences in perception

What do the results mean?

After everyone has answered the questions, gather the responses from everyone. The generation of insights step should consist of two stages, in this order:

Tally chart of responses

While we have no particular template for this, try to end up with something like this:

Tally Chart

Keep it simple here, the less setup friction the better when transferring data from the responses to the board.

Add insights

Use the chart to identify areas for discussion. Areas where there is wide agreement or disagreement or split opinion would be good places to start:

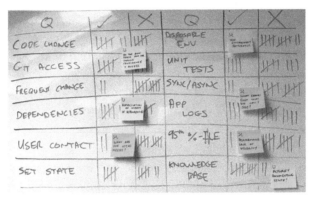

Tally Chart Analysis

Add your insights and allow some time for discussion and debate. We've used stickies to denote positive areas and potential improvements. This should set the tone for the rest of your journey, plus provide some early ideas for areas of focus.

1.3 Exercise: use Trade-Off Sliders to guide your testability focus

You may have determined a starting point but it's time to deal with a dose of reality. There are always competing priorities for a team's focus. While there may be specific areas a team wants to improve testability, this should be considered with the whole picture in mind. This is particularly prevalent early in the journey. We need to prove to those who fund our efforts that testability assists delivery. Specifically, predictability and knowledge of a system's quality. The context you are working in is incredibly important.

Gwen Diagram, in a role as part Tester and part Scrum Master at a social media startup, found that testability had to contend with many other priorities:

> "The focus was delivery, while we were trying to build not just new features but new concepts in short timescales, having deep test automation and lots of browser coverage just wasn't that important compared to being fast to market. The feedback from the product team and customers was massive, any problems we would fix forward. Testability would have to wait."

Here's a secret, though. Knowing the stakeholder priorities is a massive part of implementing testability improvements. If the priority is new features, the focus might be more on the ability to observe and control the release of the new features. If the priority is scaling, the focus might be on controlling and understanding the impact of load on existing components.

In order to generate these insights, the second exercise of this chapter is named "Testability Trade-Off Sliders."

The first step is to identify your priorities. This is an exercise for all stakeholders. Remember, the development team is also a stakeholder, which is crucial for implementing testability improvements. Your sliders might look like this to start with:

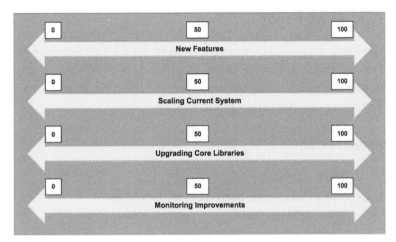

Priority Sliders

The rules for using the sliders are simple and are designed to guard against everything being top priority, and therefore nothing being top priority:

- You can score each priority from 0 to 100.
- You only have a limited amount of priority points to allocate. Each slider can be allocated 50 points. In the example above you have 200 points to spend across 4 sliders.
- Everyone has the same amount of points, regardless of the role they fulfill.

Testability is more than improving a system's logging capability or creating living documentation that describes a system. Knowing what is important to test and why is what really makes your testability effort worthwhile.

How does the exercise work?

Follow this facilitator's checklist below to help you run the exercise.

Who do we need?

We think you should throw out a wide net for this exercise; we are interested in a wide range of opinions to help expose gaps in knowledge, so invite the same people as for the Team Test for Testability exercise earlier in this chapter.

Preparation

- Time - 1 hour
- Type of Space - an open space is preferable
- Attendee Preparation
  - Ideas for what they believe are priority pieces of work
- Physical Materials
  - Same colored pens for anonymity
  - Same colored stickies or index cards for anonymity
- Digital Materials - None

Sample agenda for pre-communication

- Set the Stage (5 mins)
  - Explain Goals, Outputs and FAQ
  - Explain rules with regard to point allocation
- Gather Consensus on Priorities (20 mins)
  - Attendees to note priorities on stickies
  - Deduplicate until between 3 and 5 items remain
- Allocate Priority Points (10 mins)
  - Draw the agreed number of sliders on a whiteboard
  - Ask each attendee to allocate their points for each by writing the priority and the points allocated
  - Collect all then stick the post it notes along the sliders
- Generate Insights (20 mins)
  - Discuss the areas of perceived focus and what is expected from your testing in those areas

Facilitation

- The key point here is to let everyone have a say on what they perceive priorities to be for the team and product
- Keep it anonymous for the public part of the exercise, as priority calls can be stressful for some

Frequently Asked Questions

- When asking for priorities from the attendees, what level or size of item should I ask for?
    - Look for medium sized pieces of work. Perhaps share your product and tech roadmaps beforehand to generate ideas unless that would constrain ideas.
- Should we rank our priorities after we have determined them?
    - If three of your priorities represent 90% of the ideas, just allocate points to those.

Goals

- Build consensus on the priorities of the team, with all stakeholders in attendance
- Capture the perceptions of what's important for all stakeholders and build a picture of how aligned the team is

Outputs

- A list of de-duplicated priorities generated by the stakeholders invested in the system
- A representation of the sliders based on the de-duplicated priorities, with points allocated along the sliders according to the rules set

What do the results mean?

Discussions around priorities are often the hardest problems. You should expect the first part of the session - de-duplicating priorities - to be time-consuming. We recommend three types of priorities:

- High level & general: "rebuild system in new technology"
- Medium size & targeted: "add new video player features"
- Small & specific: "refactor the payment calculation module"

If you add them upwards targeting that medium level, then you should get a sensible list of priorities to work with. This is also true of the high-level priorities; encourage the team to break those down into the medium-sized category.

Analyzing priority point allocation

Once the team have added their points to each priority and the facilitator has mapped them out, you should get something that looks like this:

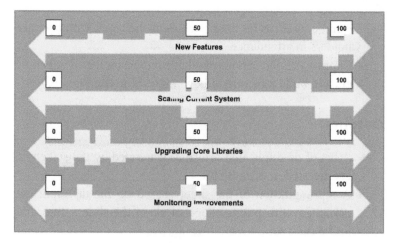

Completed Priorities

From the diagram above, you could argue that:

- The team is split on a number of priorities, apart from "Upgrading Core Libraries."
- There are stronger feelings about "Scaling Current System" than "New Features" and "Monitoring Improvements.
- There are a few outliers on "New Features" and "Monitoring Improvements" which should be investigated to surface possible misunderstandings about what the priority means..

The purpose here is to show how perceptions can differ and build some consensus over what matters to the team and where testing will provide the most value. Testability is a focusing force and this exercise should help to show the area of most impact.

1.4 Summary

Every team has competing priorities at some point. Also, there are often many improvement initiatives in progress at any one time. Testability needs be part of this wider conversation. Especially when attempting to foster a whole team approach to testing. In our experience, asking for large-scale changes for testability rarely works. Often it can divide teams further.

This chapter covered:

- All roles have a stake in improving testability but will often believe that it will not benefit them, or that it is someone else's responsibility. Advocates for testability will need to engage with these challenges.
- Performing an initial testability health check with questions about how you build software. This is inclusive for the whole team, regardless of role, starting the conversation about testability.
- Having an open conversation about where testability fits in with other system and team priorities is important. Especially

with those who fund the product. Having a balanced view of how much to focus on testability is crucial to building trust.

There is real inertia to overcome for establishing a greater focus on testability. Try to get the team thinking about testability then add it to your wider priorities. This means you will be in a strong position to make positive, iterative gains in the future.

2. Create testability targets to improve interactions with dependencies

---

Key points

- Recognize that **all disciplines within a team** can contribute to and enjoy a testable system.
- Acknowledge that your testability is **not confined to your system** and is constrained by the teams and systems around you.
- To improve our own testability it is important to **foster relationships** with internal and external teams.

This chapter covers techniques for improving the testability of software by focusing on the interacting (adjacent) services, components, libraries, and other dependencies.

2.1 Recognize needs and contributions from different roles to create an effective testability mindset

It is important to recognize that most cross functional teams have many roles within them. In our experience, individuals often identify as belonging to a particular discipline, such as testing. One of the key messages we wish to convey here is that testability is not only the domain of a tester but benefits all disciplines.

We recognize that testability should feel relevant to you in your role and have tangible benefits. Otherwise, the investment of the team and wider organization will be difficult to sustain. This guide to a few common roles may help convince those who ask questions about the value of testability and how they can contribute:

Needs	Role	Contributions
Reduce code complexity from 'workarounds' on hard to test systems. Easy to set system state and data for repeatability. Information from production on customer journeys and their relative success.	**Automation Tester**	Forthright involvement in architecture and design decisions relating to control of state. Surfacing the cost of complex automation code and data setup with precise measurements of time.

Needs	Role	Contributions
Fast feedback on change to the codebase, using tests of different levels, to expose regressions and show value. Insights and tools to diagnose problems and set systems to a known state.	**Developer**	Contributing to all levels of testing, from unit to end to end testing, improving the quality of test automation code. Having accessible and controllable application logging. Building mocks for external dependencies. Applying feature flagging to new features.
Able to express the vision for the product to the development and operations teams.	**Business Analyst**	Use analytics and customer usage data to enhance feedback for the team to act on. Informing which customer journeys represent the greatest value or risk to inform testing.
Able to learn from the current platform to make meaningful upgrades. Advocate for reliability and scalability testing.	**Ops Engineer**	Expose system diagnostics to complement customer usage information. Contribute to feature inceptions and incident post-mortems.

Needs	Role	Contributions
Release value when it's ready with a relevant conversation about quality attained versus quality needed.	**Product Owner**	Prioritize the need for testability as part of features. Advocate that testability characteristics are part of the definition of ready or done.
Maintain a long-term vision for system architecture, while delivering a roadmap with business constraints.	**Head of Technology**	Treat testability as a first-class requirement when creating and implementing a wider vision.

All improvement endeavors need advocates and testability is no different. Once those involved see the benefits, we hope they will be at the forefront, leading the change. This will leave you in a better position to tackle the next level of this challenge. The teams, systems and dependencies you interact with.

Look beyond the team for the testability impact of dependent teams and systems

How testable are the systems you depend on? Your testability is constrained by how easy it is to test a team or system you depend upon to provide customer value, which could include the systems you need to integrate with to complete a customer journey, for example. If your system relies on a payment gateway which suffers from low testability, your end to end tests may fail often. This might block your deployment pipeline, making release decisions problematic. Most systems have integrations with other, internal and external systems. Value generation occurs when those systems work together in concert. Allowing customers to achieve their goals. You have a problem if you optimize your own testability but neglect your dependencies; you are optimizing locally, at the expense of the global goal (Forte). That means you can only achieve a flow rate as large as that allowed by the most constrained bottleneck.

2.2 Exercise: employ Testability Dependency Targets to improve interactions with dependent teams and systems

Establishing that you may have a testability challenge with a dependency is one thing; determining what to do about it is another. On the one hand, you could argue that if a dependency is hard to test, it's not your problem. External dependencies might have contractual constraints for reliability, like Service Level Agreements. Contract terms and reality can be far apart; in our experience, service level agreements are not very effective agents of change. It's time to bring the team together to work on your "Testability Dependency Targets."

What does the Testability Dependency Targets exercise involve?

We need to identify who impacts your testability. This includes where they are in the organization (or outside) and how they work. You can then start to visualize the impact they have on how testable your own system is. To this end, we have identified four categories of testability dependencies to target:

Internal Systems	Internal Dev Teams	Support Teams	3rd Party Systems
Anything built by an internal team that you need in order to test your system	Any team that works on the same codebases as you, or dependent ones.	Often teams managing infrastructure or databases that you depend on are a separated function	Anything built by an external team that you need in order to test your system

We can visualize the impact of stakeholders on testability with a graded stakeholder map:

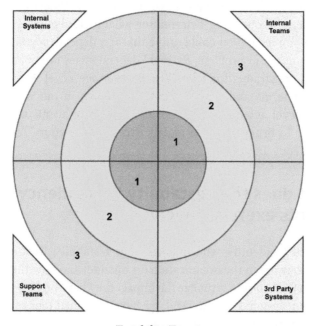

Testability Target

Lets classify each team simply so we can get a handle on the amount of impact they have:

Impact Level 1	Impact Level 2	Impact Level 3
The team or system is impenetrable to enquiry, change is unannounced and environments are routinely unavailable	The team or system can be opaque with what is being built or changed, environments tend to be unavailable on an infrequent basis	The team or system is transparent with what they are delivering, environments are available when needed 95% of the time and regularly showcases their features with supporting documentation

In order to move forward in your testability journey, it is important to understand the constraints under which you are testing your system. Gaining a holistic understanding of your relationships with other teams and systems brings your teams closer together and helps to determine common solutions.

How does the exercise work?

Follow this facilitator's checklist below to help you run the exercise.

Who do we need?

We think it would be better to have your immediate team for this exercise, as you will discuss how other teams and products are perceived. Perhaps any architects or operations focused roles would be useful, although not essential at this stage:

- testers
- developers
- analysts

Optional:

- architects
- ops

 For full details of the roles, see Team Test for Testability exercise in Chapter 1.

Preparation

- Time - 1 hour
- Type of Space - Not a boardroom, open space is preferable
- Attendee Preparation - None
- Physical Materials
    - Whiteboard or flipchart paper
    - Different colored pens
    - Sticky notes or index cards
- Digital Materials - None

Sample agenda for pre-communication

- Set the Stage (5 mins)
    - Explain Goals, Outputs, FAQ
    - Gather attendees into groups of 3 or 4
- Gather Data (10 mins)
    - Using the four categories of testability dependency targets, ask the groups to identify teams or systems and place them into categories
    - Each group produces a list, one team or system per sticky note
- Filter and Sort (10 mins)
    - Ask the groups to present their lists
    - Remove duplicates from the lists
- Generate Insights (30 mins)
    - As a whole team, from your deduplicated list, agree and place each team or system into one of the categories, on a scale of 1 to 3 on how much they impact your testability

Facilitation

- A large representation of the target above is needed
- When splitting the teams, try and mix the roles up too, to add balance to the exercise
- In order to pre-load the attendees with understanding, you could send out:
  - Four Categories of Testability Dependencies
  - The Graded Stakeholder Map

Frequently Asked Questions

- What does available infrequently mean in the impact levels?
  - Whatever is meaningful in your context. If you release every day, then something being unavailable 2 days a week is a problem
- Listing the teams and systems may seem a little repetitive but it's designed to surface a range of dependencies, even small ones

Goals

- Build consensus on the teams and systems that have an impact on your team and system
- Determine the dependent team and system interfaces to improve based on their current impact

Outputs

- A deduplicated list of dependent teams and systems, generated by consensus
- A target diagram with each dependent team or system in the relevant section, with those in the middle as a key improvement focus

What do the results mean?

The Gather Data step for the exercise should produce something that looks like this, from each team:

Internal Systems	Internal Dev Teams	Support Teams	3rd Party Systems
Authentication	Android App Team	Database Admin	Credit Checking
Billing Service	iOS App Team	User Experience	Identity Service
Load Test Data Generator	Delivery Engineering	User Interface Design Team	Payment Gateway

The groups will likely come up with many of the same answers but each group will equally likely note something the other group hasn't thought of. The example above has the "Load Test Data Generator" for example, which perhaps a tester or a devops contributor might add, but a business analyst might not.

Focusing on the target

Once you removed the duplicates from the list and asked the team to place each item from the list on the target, a picture should emerge showing clusters of responses (see image following).

There are a few factors at play in this example. Third-Party systems are often problematic, as they are products of another organization. We typically have less ability to observe, control and understand these systems. The supporting teams in the example also have a significant impact. This suggests there may be opportunities to improve relationships and understanding. From the exercise, the team can pick one or two teams or systems to focus their improvement activities. Taking another big step in their understanding of all the factors which affect their testability.

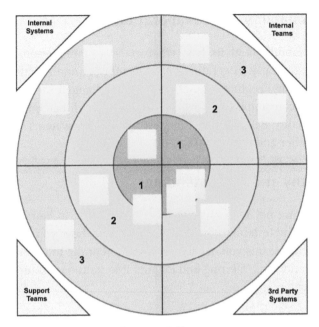

Target with Teams

Try to focus on the following areas after you have completed the exercise: Observability and information flow; Controllability and collaboration; and Empathy and understanding.

Observability and information flow

Enhance observability to provide real feedback about your interactions with dependencies. Rather than logging only your system events. Interactions with dependencies are part of a journey through your system. Both internal events and interactions should log to your application logs, exposing the full journey. Replicate this pattern for both production (live) and test environments. The key benefit: you'll provide context-rich information that the people who maintain that dependency can act upon.

Controllability and collaboration

Controllability is at its best when it is a shared resource. This encourages early integration between services and a dialogue between teams. Feature toggles for new or changed services allow for early consumption of new features without threatening current functionality. Earlier integration testing between systems addresses risks earlier and builds trust.

Empathy and understanding

Systems are not the only interfaces which need to be targeted to improve the testability of your dependencies; how you empathize with others teams you depend on needs attention, too. Consuming their monitoring, alerting and logging into your own setup helps a great deal.

2.3 Summary

Testability is a Systems Thinking (Goodman) endeavor. Concerned with feedback loops and how the structure of the organisation influences how we test. How testable the teams and systems around you are is one of key constraints you will face. Determining who and what you integrate with and how you will improve those interactions is important. It's one of the first steps to addressing your own testability problems. When you have improved the testability of your dependencies, where to focus your own efforts becomes clearer.

This chapter covered:

- Improving your own testability provides local optimization, whereas improving globally is the aim. Remember, your users and operations stakeholders are generally concerned with the whole, not only your part.

- Understanding how internal systems, third parties, support teams and internal development teams impact your testability. The links between your team and system and those nearby gives vital clues to where testability gains are.
- Taking a collaborative approach to improving testability of dependencies, keep these principles in mind:
    - Observability and information flow where the whole journey is the aim, including dependencies.
    - Controllability and collaboration to encourage early integration and risk mitigation.
    - Empathy to understand the problems and pain of those who maintain your dependencies.

As a first step, try to build a relationship with the teams on which you depend. Show them how you test and what you have built. In turn, understand how they test their challenges and how you might be able to assist. This will unlock large testability gains for you and your team.

3. Adopt testability mapping to expose hard-to-test architectures

Key points

- Low-testability architectures contribute to the conditions for shallow testing, or a neglect of testing entirely.
- Symptoms of low testability architectures are present in a team's everyday work. Problems need to be surfaced before they can be addressed.
- High-testability architectures enable different types of testing and a comprehensive view of quality.

This chapter covers techniques for detecting low testability in software architectures.

3.1 Poor architectural testability causes slow feedback and flawed decisions

In order to make changes to your architecture which will have the biggest positive impact on your testability, we need to understand what symptoms are present, their roots and how to judge progress on an ongoing basis. Without gathering data first and regularly, we are at risk of:

- Prematurely optimizing our architecture based on a single point in time rather than allowing the picture to evolve as we discover more.
- Changing our architecture with a bias to local optimization, rather than removing the overall constraints of the system.
- Focusing on our own architecture when some of our biggest testability challenges might lie within our dependencies.

Over the course of a system's life, many decisions are made along the way based on needs of stakeholders, available technologies, contributors' knowledge, and even the problems that occur during development. Often, these decisions can fundamentally influence the testability of an architecture.

> When Ash was working at a large credit broker he experienced this:
>
> *I was working on a team performing a major replatforming exercise. They were essentially creating a replica of an existing system built on LAMP (Linux, Apache, MySQL and PHP) during which the database was being ported to Microsoft SQL Server. This was mainly because the rest of the organization was Microsoft-based and had little MySQL knowledge.*

> *Doing so put the system architecture into a "bespoke" state, which no-one had seen within the organization before [which no-one within the organization had foreseen?]. Plus, support for database drivers between PHP and Microsoft SQL Server was sparse, as was knowledge of successful implementations in the wider development world.*
>
> *In addition, all the monitoring tools were subsequently split between two technology stacks, which made it hard to observe the whole system easily, as it removed the previously complete monitoring which existed for the LAMP application.*
>
> *The team dynamic changed too. The PHP development team used to support MySQL to a significant degree (as they were the only ones with sufficient knowledge to do so), but after moving to the new technology, enhancements, updates and support were assigned to a large team of Database Administrators, which limited our control of a layer of our own architecture.*

We knew we had a fundamental, structural risk within our architecture. How does a team protect the system from future impacts of such a change?

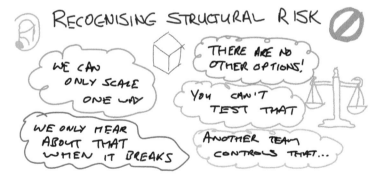

Structural Risk

You will likely have examples like this within your own history. We will go into recognizing the symptoms of poor testability in more detail later in the chapter. When faced with decisions about the architectural testability, consider the following risks:

Risk	Impact
If it's hard to test, it won't get tested	If your architecture is hard to configure, operate and analyze, testing will be too. Moreover, tests likely won't be run, or at least they will uncover less useful information
Important problems are hard to find	We want to find the problems that have the greatest impact on our stakeholders. High complexity and low observability can hide problems during testing and in operation. Production problems that are hard to reproduce in earlier environments occur often.
If it requires I/O it's going to be slow	If your tests rely on network or disk I/O then they are going to be slow and unreliable without test doubles.
Feedback slows when complexity grows	A tightly coupled architecture pushes testing into the higher layers of the system, which can lead to slow feedback and many combinations of tests to run without knowing which components are covered.

Risk	Impact
Test as close to the change as possible	When designing your system architecture provide mechanisms that allow the team to test the logic as close to source as possible. This allows for faster, more accurate defect isolation.

3.2 Identify the symptoms of poor architectural testability

The testability of your architecture is not always obvious. In our experience, you are more likely to notice the symptoms in your daily work. This can lead to a number of frustrations for the team and the stakeholders interested in the delivery of product. An architecture with poor testability limits the team in many ways.

> Ash experienced this while working at a large gaming company:
>
> *"We were testing a hybrid iOS application, a website with limited native components. Just to begin testing on a mobile device, multiple configurations had to be navigated. Pull a feature branch, build a development environment, configure relevant external feeds, join a specific WiFi network, change DNS, configure a local proxy to intercept and change certain headers and URI's. It would take a couple of hours on average to get going with testing, and that's when it all went well."*

The symptom here was that it took a long time to begin testing, which had a direct impact on the team:

- Poor quality of exploratory testing - everyone was exhausted by the time testing began! This led to a lot of limited, shallow testing, more akin to demonstrations.

- Long queues of ready to test work - team boards had long queues of tickets, waiting to be tested, in a mobile environment where time to device is critical.
- Limited test automation - writing automated tests for customer journeys that didn't yield false positives or false negatives was difficult. They fell behind the cadence of delivery, became regarded as unreliable and were removed from the build pipeline to stop them blocking releases.

It took a while to shrink the time taken to get from "ready to test" to "testing". However, once we did, the team and product stakeholders were much happier.

- Whole-team testing: since it now took less time to get our development environment onto a real device, the whole team started doing it more often, developers got feedback on real devices quickly, and key user journeys were protected with automation.
- Less context switching: shorter queues in testing columns meant developers and testers were synchronized, and were delivering as a team again.

We want to help you to recognize these symptoms so you can begin to act on them to discover your own root causes.

3.3 Exercise: Use testing smells to diagnose poor architectural testability

A testing "smell" is a symptom of a testability problem. We believe each of the smells listed below can be tamed by establishing a more testable architecture. This exercise aims to determine which testability smells impact your team and the extent of the impact on both the team as a whole and each of the individuals. The findings of this exercise will allow your team to make informed decisions about which testability smells to tackle first.

What does the exercise involve?

In this exercise we will first ask each individual in the team to rate 15 common testing smells in terms of the impact the smell has on their ability to do their job effectively. The 15 smells are listed below.

#	Testing Smell	Example Impacts
1	**Too many production issues**	Does your team feel that too many issues are escaping into production? Is your team's planned work frequently interrupted and delayed as a result of dealing with production issues?
2	**Pre-release regression cycles**	Does your team have to execute a lengthy regression test cycle before releasing? Does your team often find important issues during this regression cycle?
3	**Lack of automation & exploratory testing**	Does your team frequently check and confirm things that should be done using automation? Does your team overlook exploratory testing?

#	Testing Smell	Example Impacts
4	**Hesitance to change code**	Is your team hesitant to make small, regular code improvements for fear it will introduce undetected issues? Does your team feel uncomfortable refactoring the code even when they believe it's necessary?
5	**Testing not considered during architectural design**	Does your team neglect to involve testers in the architectural design discussions? Does your team neglect the impact on testing when making design decisions?
6	**Team constantly seeking more testers**	Does your team feel like they need to add more testers as a result of mounting workload and complexity?
7	**Too many slow user interface tests**	Does your team waste a lot of time preparing, executing and waiting for feedback from slow GUI tests, either manual or automated?
8	**Important scenarios not tested**	Does your team release the system without testing important scenarios because they are either impossible or impractical to test? Are there areas of significant risk that are not being tested?

#	Testing Smell	Example Impacts
9	Ineffective unit and integration tests	Does your team write unit tests and integration tests that often miss important problems? Does your team endeavor to continuously improve your unit and integration tests?
10	Cluttered, ineffective logging	Do your logs contain lots of errors and warnings even when the system is considered to be behaving as normal? Can team members quickly and easily isolate and debug issues using the logs?
11	Flaky nondeterministic automation	Does your team spend a large proportion of their time investigating failures, debugging and maintaining automation? Does your team re-run automation when it fails expecting it to pass the second time?
12	Tests that contain duplication & irrelevant detail	Does your team have tests that contain a lot of duplicate steps (usually setup) in order to get it in a state to perform the essential part of the test? Does your team have tests that contain lots of details that have nothing specifically to do with what you're actually trying to test?

#	Testing Smell	Example Impacts
13	**Issues are difficult to reproduce**	Does your team often encounter issues that are difficult, time consuming or impractical to reproduce either in your test environments or production?
14	**Issues are difficult to isolate & debug**	Does your team struggle to isolate and debug issues when they occur either in your test environments or production? Does it take days of investigation to find the root cause of a problem?
15	**Too much effort spent writing, maintaining and debugging automation**	Does your team rely too heavily on automation written at the UI level? Does your team test business logic through the UI?

When we talk about the impact of each of the testing smells consider the effects in terms of time, effort, your ability to deliver value and the satisfaction of the team in their work. The ratings are simple:

1. **No impact on team effectiveness**
2. **Small impact on team effectiveness**
3. **Moderate impact but rarely impacts team effectiveness**
4. **Moderate impact but often impacts team effectiveness**
5. **Large impact, almost always impacts team effectiveness**

Follow this facilitator's checklist to help you run the exercise:

Who do we need?

We suggest involving architects, developers, testers, ops and analysts.

For full details of the roles, see Team Test for Testability exercise in Chapter 1.

Preparation

- Time - One hour
- Type of Space - An open space with a whiteboard
- Attendee Preparation
  - Distribute testing smells sheet & explanations before the exercise.
- Physical Materials
  - Hard copies of the testing smells sheet for every member of the team
  - A whiteboard or flipchart
  - Sticky notes and pens (one color)
- Digital Materials - None

Sample agenda for pre-communication

- Set the Stage (5 mins)
  - Explain goals and FAQ
  - Introduction to the testing smells.
  - Introduction to scale of impact on team effectiveness.
- Gather Data (10 mins)
  - Create a matrix of the 15 testing smells and allow team members to vote using post its or some other indicator.
- Generate Insights (40 mins)
  - Explore the obvious differences in rating, particularly any outliers from the general consensus by encouraging individual contributions.
  - Capture insights from the discussion and a team consensus on the rating for each smell.
- Agree actions (5 mins)
  - The team discusses and agrees the top 3 smells they would like to address first.

Facilitation

- Change the wording of the questions for common terminology at your organization if you need to.
- The gather data step is better done anonymously, we want to learn about individual perceptions and the differences/similarities between them.
- Actively seek to drive discussion to attain shared understanding.
- If discussions about outliers result in deeper understanding of a smell, allow team members to adjust their rating.

Frequently Asked Questions

- What if someone wants to add a smell?
  - Check if it's covered already, but mostly say yes!
- What if the team can't find consensus on a rating?
  - If the team can't agree then go with the lower proposed rating. Most likely if there's a lack of consensus it's not the most important issue.

Goals

- Cultivate a shared understanding of the impact of not having a testable architecture, on individuals and also on the team as a whole.
- Create an appetite within the team to establish a testable architecture.
- Capture insights that other people within the team may not have previously considered.
- Reach consensus within the team about which testing smells need to be acted upon.

Outputs

- A matrix of smells, team consensus and interesting points.
- A list of the top 3 testing smells that the team is committed to trying to address.

After the exercise you should have something like this:

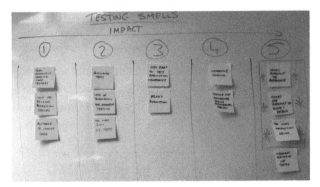

Testing Smells

What do the results mean?

As you can see each of the smells has been allocated to an impact level based on the team consensus. One could infer from the results that this team believes the following:

- They have a reasonably balanced test strategy without any significant problems in the effectiveness of their test automation, aside from a high maintenance cost of their automation due to flaky tests.
- By far their biggest concern is when issues during development are hard to reproduce, isolate and debug, which may contribute to the incidence of issues leaking into production.

The next step is to determine what this might mean for architectural testability. The team agreed to tackle three areas highlighted with asterisks in the photo above (*issues difficult to reproduce, issues are difficult to isolate/debug,* and *too many production issues*). We hope this inspires a few ideas for tackling poor architectural testability that can be refined in the subsequent exercises of the chapter, such as:

- Difficulty in reproducing issues might be mitigated by implementing controllable log levels by request or a unique ID which is attributed to a transaction throughout its journey through the system.
- Issues that are difficult to debug and isolate might be assisted by "stubbing" dependent services in order to fix the response type or response time to limit variance and narrow down possible causes.

3.4 Exercise: adopt testability mapping to measure testing feedback and waste

Typically, when we encounter hard-to-test systems there is a distinct separation between the testing performed by the developers and testers in a team. Often, the developers invest their efforts in writing unit and integration tests while the testers write UI tests and perform the vast majority of exploratory testing. Typically, the tester has little visibility or involvement in the testing performed by the developer and vice versa. As a result, the unit and integration tests (performed by the developers) rarely find important problems, while the UI automation requires constant attention (from the dedicated testers), which reduces the amount of exploratory testing done.

A testable architecture helps break this "silo culture" and promotes whole-team involvement in design discussions. By involving testers in design you bring a wealth of testing knowledge and context to the discussion. Team members work together to understand the important quality attributes, critical paths, core components and associated risks and agree on a design that allows the team to mitigate those risks in the most effective manner.

A testable architecture allows the team to build effective automation that's easy to write, fast to run and whose results can be trusted. From a developer's perspective, it provides the confidence to make small changes continuously, knowing that regressions will be detected almost immediately.

There's also the added benefit of making defect isolation easier by encouraging the creation of isolatable components which allow for pinpointing defects; this defect isolation is further helped by greater observability which allows us to determine the defective components internal state from metrics, logs and alerting (Waterhouse).

 From a tester's perspective, a testable architecture allows them to increase the depth of their testing, to perform tests that would otherwise be impossible or impractical, and to provide more detailed information to developers when issues are found.

What does this exercise involve?

As a team you will identify the different types of testing needed to deliver a product increment. This could be within whatever planning sessions you use as a team. This might be release or sprint planning, for example, or a story planning session (often known as a "Three Amigos" session (Dinwiddie)).

The key to this exercise is collaboration between product, development and test stakeholders. For each type of testing that is meaningful to you, create what we have called a 'testability map.'

You might run this exercise when you have a major decision to make regarding your system architecture such as:

- Larger new features that represent a departure from current architectural patterns, such as adding a new service as a microservice.
- The introduction of a new technology, such as a container orchestration engine or resource-intensive operations into cloud services.
- A new project or product would benefit from this thinking too. Adding functionality which uses File Input and Output for example will present testing challenges which are best anticipated and mitigated earlier.

What is a testability map?

A testability map looks like this:

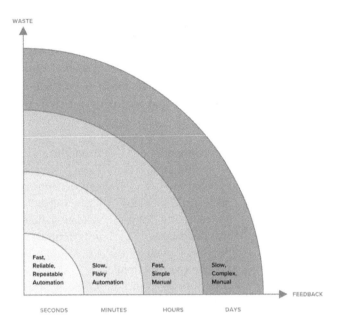

Example Testability Map

- The Y axis of the maps concern the amount of waste that occurs when attempting to test. If you have dependencies which are unreliable and suffer from significant downtime then testing may take longer or you might do more testing to gain confidence.
- The X axis concerns feedback over time, the longer it takes to receive feedback on a deployment the more risk is accumulated. For example, if you have a system with limited test automation and data dependent manual tests which require a lot of setup, then it may take days to get meaningful feedback.
- Each segment (for example - Slow, Complex, Manual) is a representation of the amount of waste you will incur in order to get the feedback you need.

You need to run the exercise for these 4 types at a minimum. if you don't perform one or more of them, then your testing lacks balance. For example, too much focus on automated acceptance testing at the expense of exploratory testing could increase your exposure to unexpected problems later in a feature's development cycle, when you could have applied variance earlier on to avoid them. You can of course add more testing types that are important/challenging for you:

Testing Type	Description	Examples
Acceptance Testing	Demonstrates that the system can successfully provide the capability that it was designed to provide. These tests are often derived from user story acceptance criteria or BDD scenarios and aim to confirm the feature works as designed	Basic flows, sunny day scenarios, business logic, validation, functional tests, contract tests.
Exploratory Testing	Exploring, experimenting and interacting with the system in varied ways in order to unearth the unknown and unexpected. This type of testing is focused on finding problems and developing a deeper understanding of the system.	User journeys driven by personas, resilience, error handling, failure scenarios, usability, rainy-day scenarios, risk-based tests, stress tests.

Testing Type	Description	Examples
Regression Testing	Detects unintended behavioral changes in the existing functionality that have been introduced as a result of adding new functionality. Typically, acceptance tests become a part of your regression test suite once a feature has been released.	Basic flows, sunny day scenarios, business logic, validation, functional tests, contract tests.
Operability Testing	Detects and isolates issues when the code is deployed in a fully functional pre-production or production environment. These typically focus on ensuring that business critical functionality is working when deployed in a given environment.	End to End Synthetics, User journey tests, monitoring, alerting, logging, analytics, instrumentation, performance testing.

How does the exercise work?

Follow this facilitator's checklist below to help you run the exercise.

Who do we need?

We suggest involving architects, developers, testers, ops and analysts.

For full details of the roles, see Team Test for Testability exercise in Chapter 1.

Preparation

- Time
  - 1 hour - as part of planning a product increment or architectural change.
- Type of Space
  - An open space with a large whiteboard
- Attendee Preparation
  - Familiarity with the current system architecture and the proposed changes.
  - Familiarity with the four testing types:, acceptance, regression, exploratory and operability.
- Physical Materials
  - A whiteboard or flipchart
  - Sticky notes and pens
- Digital Materials - None

Sample agenda for pre-communication

- Set the Stage (5 mins)
  - Goals, Outputs, FAQ
- Draw the architecture (20 mins)
  - A member of the team draws the system architecture with assistance and feedback from the team.
  - The facilitator needs to timebox this exercise and ensure that the discussion stays focused on the topic.
- Associate Test Ideas with the System Architecture (20 mins)
  - Get the team to brainstorm test ideas, adding a sticky note for each test idea they believe is necessary in order to release the product increment with confidence.
- Position stickies on the testability maps (20 mins)
  - The team works through each of the segments placing each of the stickies in the appropriate area.

Facilitation

- Ask for a volunteer from the team to draw the system architecture with assistance & feedback from all members.
- Run through each question & seek clarity on each of the answers until you feel the team is happy to move on.
- Take a picture of the system architecture (in case you have to remove it).
- Draw four testability maps on the whiteboard labelled acceptance, regression, exploratory and operability.
- Explain which test ideas belong in each of the segments.
- Prompt the whole team to add stickies to the appropriate segment for each testing idea or task they believe is necessary to deliver the feature with confidence.
- In order to help generate test ideas, consider using a testing mnemonic as a framework (McKee). This will help the test idea generation to flow, or at least get started.

Frequently Asked Questions

- What if the team can't agree on the architecture?
  - Allow each member of the team express their views; if the team can't agree, the team lead decides.

Goals

- Create a shared understanding in the team of how the current design may inhibit efficient testing.

Outputs

- A set of testability maps which express the types of testing most meaningful to your team for that feature, product increment or architectural change.

After the exercise you should have something like this:

Example Testability Map - with details

What do the results mean?

This team were focused on regression testing testability map, as they were looking to upgrade a number of their dependencies, specifically their versions of PHP using to build their HTTP API, Website and Batch File Processing system, plus MySQL for data storage.

- **Seconds** - The team recognized that they had a number of tests for their HTTP API, which used real internal dependencies and mocked external dependencies. These would be very useful for fast feedback on PHP version changes in their API, but they had no fast mechanism for getting feedback on their website.
- **Minutes** - The team's automated user interface tests sat in this segment, and the website received less attention as API traffic increased so the risk of upgrading versions felt greater. As the PHP version changes and the code changes for deprecated

methods were made, these would be better (at least partially) be run in their development environment.

- **Hours** - the team felt that database integration tests for crucial information for billing customers and audit information for compliance purposes would be needed. In addition, the team wanted to do some testing of their API with the real version of the payment gateway, rather than the mocked version in their development environment. They decided to bring these external card gateway tests into the 'Minutes' segment, but run them on a more infrequent basis in their pipeline.
- **Days** - The pattern you can see in the testability map for regression testing is fairly common, and the team that completed the exercise knew their major problems lay in the less frequently run monthly reporting and batch jobs. Although they run less frequently, they are still critical functionality but have very slow feedback; plus, they are only set up in test environments closer to Production. The team agreed that testing these in earlier test environments would be prudent.

3.5 Summary

This chapter covered:

- The risks of low architectural testability, which we believe not only compromises your ability to test effectively but also your overall ability to deliver.
- Listing the common smells of low architectural testability and which ones matter the most to the team currently. This shows where your pain lies in the team's perception. Problems oftem get underrated as time goes by and familiarity increases.
- Diagnosing how your current architecture and future changes affect the effectiveness of your acceptance, exploratory, regression and operability testing.

Low architectural testability is a real source of pain for teams. We often inherit systems we don't understand from previous teams, then we do our best to deliver with a hard-to-test system. This chapter has expressed that the key is to detect and diagnose architectural challenges to effective testing.

4. Apply the CODS model to increase architectural testability for faster feedback

Key points

- Striving for testable architecture **breaks down silos** within software development by involving testers in design discussions from the beginning
- A testable architecture needs a balance of **observability, controllability, simplicity and decomposability**
- Aim to **build in testability early**: retrofitting testability is much more difficult
- Isolate **dependencies** to control their influence on testability
- The benefits of greater testability extend **beyond the development team**, translating into a product which is easier to operate and support

This chapter covers techniques for improving the testability of software architectures.

4.1 Explicitly design your architecture for testability

Establishing a testable architecture means designing the system to make it inherently easier to test. It allows teams to find and fix important issues faster, make changes with greater confidence and ultimately deliver software according to the needs of your stakeholders.

We should say here that creating a testable architecture isn't an easy task. Many teams work on systems that they have inherited and have been operating in production for some time. Other teams work on new products with high expectations for delivering features at scale. We have designed the chapter to address both.

However, when a team focuses on designing for testability it allows members of all disciplines to experience the following benefits:

- Faster development through fast, reliable feedback in the form of unit level tests.
- Teams begin testing earlier as the system is decomposed into small, testable chunks.
- Test interfaces provide a mechanism which allows the team to do broader and deeper testing.
- Well-defined system boundaries and test doubles allow for more robust automation.
- Greater controllability and observability ultimately lead to faster defect isolation, debugging and resolution of customer issues.

In order to realize the above, this chapter will focus on:

- Creating a compelling vision of what characterizes a testable architecture. We believe it is important to describe what testable architecture looks like so you can then describe it to your stakeholders.

- Perform comprehensive reviews of the testability of your architecture using the principles of decomposability, controllability, observability and simplicity.
- A case study of how a team improved their ability to safely deliver value by designing their architecture for testability.

4.2 Principles of implementing high testability architectures

This chapter will address practical ways to improve the testability of your architecture, but we believe that these practical exercises should be underpinned by certain principles. Without these principles in mind, any inferences made from the exercises might result in a shallow understanding of the true aims of architectural testability.

Embracing constant change

Achieving a cadence of change that meets stakeholder needs requires timely feedback on risk and impact. Establishing a testable architecture allows your team to create automated test suites that provide valuable, comprehensive feedback ideally within minutes, if not seconds. Ideally, each team member can run a comprehensive suite of tests every time they make a change. This feedback provides a critical safety net that encourages your team members to make changes frequently without fear of introducing regressions. The confidence that your team gleans from these tests encourages good behavior in the form of refactoring and continuous improvement of the code base. This means that a testable architecture allows your team to make changes with very low cost and risk.

Driving better design

When we encounter a system that is hard to test, it's almost always as a result of design problems. By addressing these design problems we not alone get a better design but also a more testable system. When we consider why, what and how we will test during design it helps us to decompose the system into independent, understandable chunks with clearly defined responsibilities. Therefore, when we design for testability we make decisions that result in a more decoupled, cohesive architecture.

Minimizing waste

Making testing a primary concern during our design discussions is a hugely important ingredient in establishing a testable architecture. Trying to retrofit testability is often impractical. However, designing for testability has negligible development cost while demonstrating substantial benefits in terms of eliminating waste. A testable architecture allows the team to build small, independently testable pieces of work meaning that we can minimize rework and begin finding problems much earlier in the development cycle.

A whole-team focus on testing, especially during design discussions, provides a mechanism to raise important questions and concerns that may impact our ability to test effectively and efficiently. This process in itself often helps identify design problems such as tight coupling between components or unreliable external services that would otherwise have been overlooked.

Developing better customer relationships

Testable architecture allows the team to test the system in deeper, broader ways by exploring risks that might otherwise be impossible or impractical to test. Testable architecture provides a means of injecting faults, simulating unreliable resources and easily putting the system under test in the required state. This means the team can now find important problems before they impact the customer. When issues occur in production, testable architecture also allows the team to identify, isolate and fix the failure quickly, therefore minimizing the negative impact on the customer.

Working at a sustainable pace

Testability is integral to the flow of everything in software development, it allows the team to build reliable automation that provides valuable feedback on every change the team makes. This automation provides a foundation that allows developers to continuously improve the code. It also allows testers to continuously explore and interrogate the system in new and interesting ways. This approach frees the team to do rewarding work at a sustainable pace while still delivering valuable software to the customer.

4.3 Exercise: Use 'CODS' to increase architectural testability

Typically when we encounter hard-to-test systems there is a distinct separation between the testing performed by the developers and testers in a team. Often the developers invest their efforts in writing unit and integration tests while the testers write user interface tests and perform the vast majority of exploratory testing. Typically the tester has little visibility or involvement in the testing performed by the developer and vice versa. What results is a situation where the unit and integration tests rarely find important problems while the user interface automation requires constant attention and the tester rarely has time for exploratory testing.

A testable architecture helps break this silo culture and promotes whole-team involvement in design discussions:

- By involving testers in design you bring a wealth of testing knowledge and context to the discussion.
- Team members work together to understand the important quality attributes, components and risks and agree on a design that allows the team to mitigate those risks effectively.

- Having a shared understanding of your team's testing goals allows each of the team members to align their efforts to help the whole team succeed.
- It provides the confidence to make small changes continuously, knowing that regressions will be detected almost immediately.
- A testable architecture allows them to increase the depth of their testing, to perform tests that would otherwise be impractical, and to provide more detailed information to developers when issues are found.

What does this exercise involve?

This exercise involves identifying potential architectural improvements to enhance testability using four attributes: Controllability, Observability, Decomposability, and Simplicity.

Attribute	Description
Controllability	In order to be able to test effectively we need to be able to identify and control the critical variables in the environment that influence how the system behaves. Controllability provides the ability to control the system in order to visit each of the system's important states.
Observability	Observability is our ability to see everything important in the system. When testing we need to be able to see what's happening in the system to determine where problems may be occurring.
Decomposability	Decomposability is the ability to divide the system into independently testable components. When testing we need to be able to break the system down into components that can be tested in isolation. As long as the component respects its contract with the other dependent components in the system there should be no need to retest the other components.

Attribute	Description
Simplicity	Simplicity is how easy the system is to understand. The more inputs, moving parts and interactions in the system the more opportunities there are for failure, resulting in a system that's more difficult to test.

Each of the attributes has a number of potential strategies associated with it, here are some examples:

Controllability

- Feature flagging techniques to control exposure of new technology or functionality
- Canary releasing to expose new technology or functionality to a limited number of consumers
- Automated deployment of the application to be able to deploy new versions on demand
- Blue/green deployments to switch between new and old versions to protect current and new functionality
- Create test environments which mirror the production environment to a high degree

Observability

- Logging in each layer of the architecture and between components
- Controllable levels of logging by request to gather further information on consumer requests
- Aggregated logging to bring together logs from various layers and interfaces to show holistic pictures of systems events
- Ability to replay traffic to observe events in the architecture
- Monitoring combining diagnostics (CPU usage for example) and application information

Decomposability

- Employing a microservice architecture to decouple critical components particularly when transitioning from a monolithic architecture
- Creating circuit breakers between layers to handle persistent error conditions between internal and external services
- Adding queueing technology such as Kafka to manage high-load scenarios or to queue infrequent operations to be processed at a different time
- Creating mocks for external services to isolate parts of the architecture for testing and issue diagnosis

Simplicity

- Automated API documentation tooling such as Swagger
- Developers and testers supporting the system in Production
- Policies to pay back and discourage build-up of technical debt such as test-driven development
- Pair and mob programming to promote knowledge sharing between developers and operations people
- Adopting Lightweight Architecture Decision Records in source control to track changes to the architecture (Henderson)

How does the exercise work?

Follow this facilitator's checklist below to help you run the exercise.

Who do we need?

We suggest involving architects, developers, testers, operators and analysts.

 For full details of the roles, see Team Test for Testability exercise in Chapter 1.

Preparation

- Time - 2 hours
- Type of space - An open space with a large whiteboard
- Attendee preparation
  - Familiar with the proposed or current architectural design.
  - Familiar with the above descriptions of decomposability, controllability, observability and simplicity.
- Physical materials
  - A whiteboard or flipchart.
  - Sticky notes and pens (one color).
- Digital materials - None

Sample agenda for pre-communication

- Set the stage (20 mins)
  - Describe the goals, outputs, FAQ
  - Draw a representation architecture on the whiteboard
- Review the design for testability (50 mins)
  - Take the following inputs and map them to the architecture, based on where their impact was felt the most:
    * Any high priority, customer-impacting production incidents that have occurred in the last 6 months.
    * Any team-impacting incidents, such as environmental downtime or internal networking problems.
    * Bottlenecks and constraints identified during performance or capacity testing.
    * Analysis of where recent changes to the code base have occurred. Static source control analysis tools such as Gource can help to visualize where change happens often.
    * Any new features or improvements that may be in progress or coming up on the roadmap.

- Recommend improvements (50 mins)
  - Review the architecture and look for opportunities to improve the decomposability, controllability, observability and simplicity.
  - Add potential opportunities to improve onto the architecture you have drawn.

Facilitation

- Ask for a volunteer from the team to draw the system architecture with assistance & feedback from all members, leave plenty of physical room around it as you will be adding to it.
- Run through each topic and seek clarity on each of the answers until you feel the team is happy to move on.
- Take a picture of the system architecture (in case you have to remove it).
- Encourage the team to explore ways of using the four testability attributes (decomposability, controllability, observability and simplicity) return to the definitions if needed to restart the conversation.
- Be aware that a balance of the four attributes of architectural testability is the goal; if a team focuses solely on observability (for example), reintroduce the other attributes.

Frequently Asked Questions

- What if the team can't agree on the architecture?
  - Allow each member of the team express their reservations and if the team can't agree the team lead makes the call.
- When does the team know when they are done?
  - When the team has an agreed representation of the architecture, located areas for improvement using data available to them and identified improvements using the attributes of a testable architecture.

Goals

- Create a shared understanding in the team of how the current design may inhibit efficient testing.
- Expose testability limitations in the design using available historical data and anticipated changes.
- Rework the design to mitigate the testability limitations.

Outputs

- An architectural design that satisfies the team's testing needs in a way that facilitates fast feedback with minimal waste.

After the exercise you should have something like this:

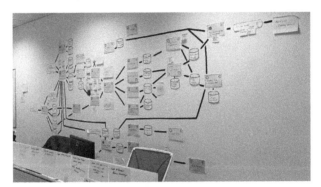

Example Exercise Output

What do the results mean?

The team that completed the exercise above came up with a detailed plan of their architecture, then mapped the types of testing they did to recent customer-impacting issues. This was a good opportunity to get feedback from both development and production environments woven into the design.

After discussing and mapping the current architecture, three main themes occurred to them:

- Changes to a networking layer that depended heavily on a centralized team (with which they had little communication) affected more than their own application and caused a poor working relationship with that centralized team.
- The team felt it was often hard to set up test data for a number of minor functional problems in the user interface.
- There were more serious, but slower to manifest, problems in the application layer around performance and memory usage, particularly when retrieving data from multiple resources (including external dependencies) to provide a coherent result.

For each of the problems, we picked out an improvement using the four principles of architectural testability. A few examples are in the table below:

Problem	Principle	Recommendation
Network Layer	Simplicity	Receiving the alerts the centralized networks teams received for traffic pertaining to their application, thus understanding the challenges better, with a view to understanding it in future.
Test Data	Decomposability	Creating a service over HTTP between the problematic user interface components to transport the relevant data, decoupling interface and downstream data and external dependencies.

Problem	Principle	Recommendation
App Layer Performance	Observability	Making 'active' database threads, 'waiting' database threads, 'operations initiated' and 'operations completed' visible via a dashboard across key datastores for further analysis in two weeks.
App Layer Performance	Controllability	The main decision was to isolate the external dependencies with a set of circuit breakers (Fowler2014a), which could be opened to test with and without the external dependencies to begin to diagnose performance bottlenecks.

It's important to consider the size of the changes recommended at this point. For the app layer performance the ultimate goal would be to have data in more sensible contexts (rather than retrieved from multiple locations), thus having a more observable and controllable system. Large changes like this arenâ€™t always possible initially, and may depend on priorities and further analysis. We advocate looking for small improvements on a consistent basis to enhance testability; these build momentum for the larger changes, if indeed they are needed after a period of iterative improvements.

4.4 Moving from hard-to-test to high architectural testability

We've presented practical exercises that can help to improve architectural testability. In order to bring the themes in the chapter together, here is a case study from Rob's past experience of the challenges of low architectural testability and how teams can overcome them.

Rob writes:

A number of years ago I was hired as a test automation engineer by a young, rapidly growing company that had decided to establish a new engineering department outside of their headquarters in the US. The company was predominantly a hardware company but had developed a proof-of-concept software product for configuring and monitoring their hardware that was gaining traction in the market.

When I started, there were approximately 35 members in our engineering department including management, developers, testers, Scrum Masters, technical writers, and support specialists. As part of our onboarding, the entire engineering department was asked to help in completing the regression testing of the forthcoming release. This request served two purposes: first, to get the new recruits familiar with the product; and second, to alleviate the testing burden on the existing team.

During the next seven weeks, the entire engineering department was bogged down in torturous regression testing and hit wave after wave of testability pain including:

- *Flaky automation: The existing test automation suites were failing every day, not because they were detecting issues but because the tests were dependent on the US-based hardware being accessible and in a fully functional state.*

- *Manual inspection: As a result, we had to check thousands of attributes manually through the product's REST API. This exercise was error-prone, time-consuming and completely soul-destroying.*
- *Limited coverage: Many core test scenarios were simply impossible or impractical to test as our team didn't have direct access to the hardware, and basic lab requests took days to get completed.*
- *Tightly-coupled code: The code was very tightly coupled meaning that every time we found and fixed an issue, three or four seemingly unrelated bugs would appear, meaning that every change required a mini regression cycle across the whole product.*

After the software was released, the pain didn't abate; within days we had to deploy a critical patch and the feedback from our customers was underwhelming.

What happened next was the catalyst for a monumental change in the way we designed and tested software. The newly-appointed architect who had just endured seven weeks of regression testing came to my desk and asked me "how can we make this product easier to test?". I didn't have an answer to his question so we gathered one of the scrum teams and went into a meeting room to discuss it further. What we decided to do was:

- *Identify the biggest testing problems (smells) we had encountered during the regression cycle.*
- *Discuss what we believed to be the root cause of these testing problems (smells).*

After having this discussion, we determined that the majority of our testing problems were caused by three factors:

1. *Our testing was entirely dependent on hardware over which we had no control. We could not control the health, availability or state of the test hardware.*

2. *We had no way to see when and where issues were occuring in the system; we could only identify problems once they had propagated through the entire system.*
3. *Every single change had the potential to break completely unrelated functionality in the system.*

Having identified the root causes of our testing difficulties we decided to create a testing wishlist, a list of things that if implemented would eliminate our 3 biggest testing problems:

1. *Provide greater visibility to the inner workings of the system.*
2. *Remove our Hardware Dependency.*
3. *Break the system into independently testable chunks.*

Soon after completing the wishlist we began our first substantial chunk of development work. We were tasked with adding support for a new type of storage array. With our wishlist in place we set about creating a design that fulfilled each of the wishes.

Provide greater visibility into the inner workings of the system

In order to provide greater visibility we decided that we needed to capture all interactions with the hardware component before we performed any processing. We would now be able to configure the system to dump the raw data captured onto the file system in either XML or JSON format.

This meant that we could now see exactly what data was coming from the hardware component, whether it was structured the way we expected, whether specific data was present, what the specific values were etc. We also made a number of changes to our logging to ensure we were capturing all critical transactions in concise, understandable language throughout the tech stack.

Remove our hardware dependency

To remove our hardware dependency we decided that we needed a realistic way to simulate all the interactions with the hardware component. We decided that if we could inject the raw data that represented the interactions with the hardware at the very lowest layer in our tech stack this would fulfill our needs. We combined the idea of being able to dump and inject the raw data gathered from the hardware component into a record and playback feature.

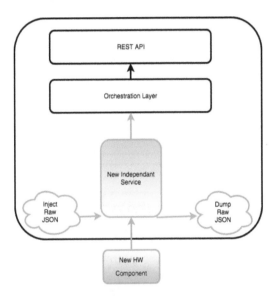

New Hardware Component

This testability feature allowed the team to configure the system to dump all the raw files onto the file system, store these files and when needed use these same files to simulate the interaction with the real hardware. This represented a huge breakthrough as we could now test the system when the hardware was either unavailable or in a bad state. An additional benefit of this new design was that we could now manipulate the dumped raw files to simulate whatever state or scenario we wished to test. We could now test things that were either impossible or impractical to test using our old approach by just editing a simple text file.

Break the system into independently-testable chunks

In order to make our new hardware component independently testable we decided that we would create it as a completely independent service that was responsible for managing all interactions with the hardware component. Our new service was exposed as an interface to the orchestration layer which meant it was decoupled from the existing code. This meant that as long as we satisfied the contract required by the orchestration layer there was no way code changes in our new service could break the existing code and vice versa. We had the added benefit of being able to deploy the new service independently of existing code.

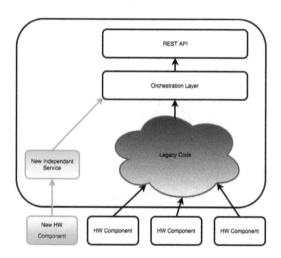

Decomposed Architecture

The results of our change

Initially there was a large learning curve for the entire team. Developers had to figure out how to implement the record and playback feature, build and deploy an independent service, testers had to test these new mechanisms while the automation engineers had to build the infrastructure to support capturing the raw logs, modifying them and injecting them. By the time we had completed the component work we were two sprints behind schedule and management were beginning to express concern about this new approach. The team however, felt that we were moving in the right direction, so we persisted, and this is when we really began to reap the rewards of our new design approach.

Earlier testing

As soon as we began work on support for a second hardware component, we found unexpected benefits to our focus on testability. The second component was completely new and was not yet physically available from the manufacturer, but we were able to get the communication specification from the manufacturer and build the raw json file that simulated the hardware interaction. This allowed us to complete the bulk of the development, testing and automation effort before the hardware was even available.

Faster development and testing

We also found that as soon as we began work on the second component, we were able to reuse much of the work that we had done for the first. The developers and automation engineers could very quickly extend their existing code and infrastructure to support the second component, and the testers were now very familiar with how to use the record and playback mechanism. This meant that we completed development, testing and automation in a fraction of the time it took for the first component.

Broader and deeper testing

For the release of the second component, we needed to be able to support a maximum configuration of 144 nodes; however, we could only physically access three nodes. Using the record and playback mechanism we dumped the raw json files onto the file system and manipulated the files to simulate 144 nodes and injected these files into the service. This allowed the team to test something that was physically impossible to test but represented an important scenario to explore.

Another benefit was that we could now record and store raw json files that represented how the hardware component interacted for a specific hardware firmware version. If the team needed to perform tests against different firmware versions for backward compatibility reasons, we could now simply inject the appropriate json files into the service, which was quick and easy to automate. To manually upgrade or downgrade a hardware component used to take days to complete and therefore this represented a huge saving in terms of time and effort.

Faster, more reliable feedback

With the record and playback mechanism in place we had managed to build a rigorous, robust and reliable automation test suite that was a core part of our Continuous Integration environment. Every morning when we arrived at work we could immediately determine if an important issue had been introduced and where.

More exploratory testing

With the benefit of rigorous, robust and reliable automation and the flexibility of the record and playback mechanism, our testers were now free to explore and test the system in ways that they had never had the opportunity to do before. Testers were now doing things like security testing, load testing and resilience testing, identifying important problems that would never have been identified before.

Better customer support

A customer support engineer approached me one day; one of our customers was experiencing a problem with the first hardware component our team had added support for. I went to his desk, we accessed the customer's environment and I asked the support engineer to put our service in record mode. We captured a dump of the raw files, copied them to our test environment and injected them into the service locally. We managed to reproduce the issue in minutes, fix it and retest it within an hour and get the issue patched for the customer in the same day. This process would have taken days if not weeks before we implemented our design changes.

Relatively quickly, news of this new approach and its success spread not alone across our office but to the headquarters in the US. The approach was adopted throughout the entire software development organization and within 12 months the 7-week regression cycle was a thing of the past. Releases now involved a handful of test engineers performing a couple of days of exploratory testing prior to release. We as a development organization felt like we were doing valuable work, at a sustainable pace, that was helping our customers solve real problems.

4.5 Summary

This chapter covered:

- Principles of highly testable architecture to guide discussions around changes to your overall architecture, as part of building in testability, rather than retrofitting it later.
- Using the principles of simplicity, observability, controllability and decomposibility in a practical way to identify where, when and how to intervene to enhance architectural testability.

- A comprehensive case study on how different disciplines can come together to overcome poor testability and design a testable architecture which helped them to deliver safely and when it mattered.

Designing for testable architecture needs a mindset shift, similar to designing for better operability. We believe designing a more testable architecture can have significant benefits to your team and organization, engaging all team members in the activity of testing, and improving team cohesion. Ultimately, highly testable architectures help you to deliver new features to customers when those features are wanted, protect current functionality, and help make supporting applications much more humane.

5. Adopt ephemeral development environments for fast feedback

Key points

- Effective use of environments allows **testing to start immediately** without delays to feedback
- Relying on **long-lived persistent test environments** often leads to apathy through lack of ownership and maintenance
- Development environments need to **evolve over time and embrace new technology** to be able to perform testing with realistic dependencies as early as possible.
- Build a development environment capable of **performing a varied set of testing activities** to increase testing capability

This chapter covers techniques for improving the testability of software through a focus on short-lived, automatically recreatable test environments.

5.1 Common challenges with static test environments

Test environments have always been a contentious topic within software development. As testers of some experience, we have been involved in many long conversations about creating, operating and maintaining test environments.

For a long time, persistent, long-lived test environments were almost ubiquitous, with all the maintenance time and costs associated. Sometimes with entire teams looking after them. Other times no one was responsible for the environments. As build automation and containerization have grown in popularity, repeatability and reliability have been enhanced. However, the most common model that we have seen is a mix of both the disposable and the persistent.

In order to get a holistic view of quality you should perform many different types of testing. To achieve this, your team will need multiple environments which resemble your production environment. Ephemeral environments are a foundation for the varied test approach advocated by testability.

This chapter is based on the premise that more testability gains can be had from enhancing your use of a local development environment, rather than persistent test environments. That is not to say there is no value in other environments, but there is a lot of guidance already about persistent test environments. We suggest you start with the book "Continuous Delivery" by Jez Humble and Dave Farley, which has excellent coverage of using test environments to perform a varied set of testing activities.

Many of us in software development will have a number of stories about issues with environments blocking a team's desire to deliver change. Ash recalls his first experience of testing a web application, after mainly working on desktop applications.

Ash writes:

"It was a big shock. I was used to entirely self contained applications. Test environments were not even part of my paradigm of software. The web application I started working on had environments that would be unavailable for days or even weeks at a time. Downtime was both planned (such as deployments, dependency upgrades) and unplanned (disk space, data corruption and those who had the relevant knowledge were unavailable to fix). I thought this was normal, as it was my only experience."

Recognize common challenges and their testability impact

The example above is indicative (although perhaps extreme) but this has persisted throughout our careers. Some problems with environments are very obvious and present, some are more subtle. The one pervasive constant is that many problems with environments are accepted as the norm, often requiring a large undertaking to fix. So we do the best we can and ignore the problems.

In order to recognize and explain the impacts on testability (and therefore the flow of information and throughput of work for your team) we have compiled what we believe to be a common (but not complete) list of problems:

Problem	Testability Impact
Earlier environments are neglected, only production or staging (in the best case) are trusted.	Testing happens later and later in the development life cycle, meaning important problems that threaten value are found too late.
No one is willing to maintain or administer shared environments, leading to confusion over who is responsible for what.	The gap between test environments and production grows wider, meaning your testing becomes less realistic.
Apathy in maintaining environments increases over time, decreasing the desire to keep the application, software and hardware dependencies production-like.	If an environment is long-lived, those who maintain it will become apathetic to it. Those who originally built it will move on and the finer details of maintenance might get lost, resulting in environments that differ from production to an unknown degree.
The ability to generate test data or safely use realistic data is limited or not present.	If you are not able to control the state of your data and therefore your application, how effective can your testing be? Data with greater realism helps to expose your system's resilience too, thus enhancing controllability.
Environments are not available when you need them, or you are running too many environments that are under utilized.	Under- or over-utilization means that you are wasting time and money, either with extra capacity that you don't use, or because you are waiting to test. Waiting in queues is an excellent indicator of low testability.

Problem	Testability Impact
Shared environments lead to limited resilience testing, as someone else always needs the environment.	Resilience and destructive testing tests the observability of a system and its ability to recover. Such testing becomes hard to schedule when there is contention for an environment.
Failing tests are ignored on certain environments as they are viewed as unreliable.	Failing tests due to environmental inconsistencies potentially obscure application failures and your ability to see threats to value before production release.
Fundamental differences with your production environment, particularly unknown differences which cause hard-to-isolate issues.	Different versions of dependencies and hardware, separate logging and monitoring policies all make the production and test environments further apart, rather than a reliable source of information for each other.
Lacking environments suitable for performance and load testing, preventing scaled versions of this type of testing in your pipeline.	Breaking the cycle of late performance testing (or none at all) is a key benefit of testability. If you cannot scale performance and load tests for early environments with trusted results, you are likely to find performance and capacity problems too late to fix in a cost-effective manner.

Hopefully these guidelines above will help you build your case for short-lived, isolated environments that are as similar to production as possible. A word of warning when you come to challenge your current test environment policy, though: shared test environments invoke strong feelings of sunk cost at times. Be prepared for people to not let go of them, as so much effort has been invested already.

5.2 Use your development environment for fast feedback

Being able to build and destroy a version of your application and its dependencies with configuration of your choosing is one of the key testability gains in terms of effective use of environments. Developers have long been working with local versions of their code with the ability to run the whole application for feedback. Having access to this as a tester - or indeed as any discipline within a cross functional team - is a powerful collaboration tool.

Utilize new technology to enhance testability in your development environment

One of the most important technologies for evolving your development environment is virtualization and containerization. This technology has enabled development teams to run complex and realistic environments locally, often including dependent systems. When Ash was working at a major ecommerce company, the development environment went through many evolutions as technology progressed.

Ash writes:

At first we had a number of services that made up the application running directly on our local machines. These weren't running on the target operating system, so we switched to a Linux virtual machine using Vagrant to build it, adding more realism to the testing done. There were two downsides to this change: memory was being permanently set aside for the VM, causing slowness on the local machine, which got worse as the application grew; and we had to run each service via Bash aliases, which was easy to forget, leading to lots of false starts in testing.

The next step was to containerize each of the services and databases, so we used Chef and Doctor to build each one. Because of the dependencies, it took a while to build a clean slate once each container had been torn down. The network traffic from downloading dependencies to our local machines and onto a VM led to many wasted afternoons. After we realized this, we created a job to build base container images in a separate repository from which local machines could pull. We also prioritized timeliness (keeping the team moving together with less context switching) and removed Vagrant from the development stack, which led to a lot more speed and freedom.

Once this container structure was in place, we could really begin to expand our testability and add some really novel features:

- The ability to share our development environment by putting a URL on the company DNS that resolves to a local machine IP address. This allowed multiple team members to test against a single environment, which was useful when testing for concurrency problems or sharing new features with stakeholders.
- The ability to build groups of context-specific containers. Testing the mobile site required only certain containers be a representative environment, and the tear-down and rebuild process took seconds. The wait for feedback was therefore minimized.
- We added bash aliases to expose application logs from each running container, and eventually even a combined application log from all running containers. Exposing logs is a basic testability enhancement often forgotten.

Now they were easy to access without needing to log into the container itself.

- *An identical environment for everyone, built from common Chef recipes. When there was a problem, or a clarification needed, each team member could build an environment with the same configuration. Therefore, replicating problems became easier and more reliable.*
- *Hooks in the development environment to other environments later in the pipeline. We would test on a feature branch in our development environment then promote a release branch to an integration environment. If a problem was found during integration, we could synchronize data in the integration environment with our development environment to replicate that problem more easily.*
- *A container with a reverse proxy to act as a load balancer. We could even complete resilience tests by throttling network traffic, dropping connectivity or removing running containers. This enabled us to judge the application's handling of disruption. Essentially, we introduced operability testing much earlier in the application's development cycle.*

This took about two years to realize fully. It certainly didn't happen overnight. There was a lot of experimentation and waiting for dependencies to download. We were lucky enough to have the managerial backing to work on this regularly, inspecting and adapting to changing architectures and business needs.

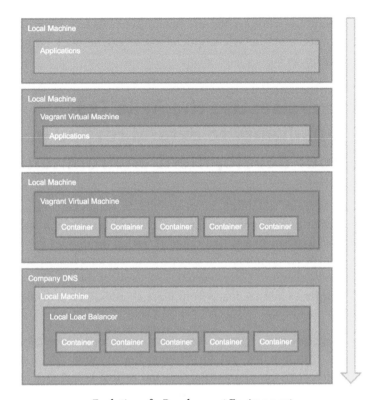

Evolution of a Development Environment

It's worth noting that in the ecommerce example above, further testing in environments subsequent to the team's development environment was still required, as they stepped through environments closer to the setup of production. For example, large scale performance and capacity testing was executed in a pre-production environment and security testing was carried out in a staging environment by a third party. The team minimized the surprises found by these subsequent testing activities with faster feedback via their enhanced development environment.

The key message here is that the team collaborated a great deal, learning about containers, orchestration, virtualization and infrastructure as code. This led the team to collaborate closely with the infrastructure, operations and site reliability teams as well, as they needed to reach out for relevant database access, network changes, and other needs.

5.3 Exercise: Use the Agile Test Quadrants to extend testing in your development environment

Enhancing your testability is an enabler to achieving balanced testing as early as possible in the development life cycle. Imagine being able to perform a varied range of testing techniques in your development environment, on feature branches or even on small incremental changes. Early warnings on security problems, page rendering times and feedback from users should all be possible as well.

Many approaches to testing overemphasize deterministic testing methods, thus missing out on the benefits of randomness and serendipity. For example, writing only automated acceptance tests and not performing any exploratory testing can miss the subtle bugs and hidden requirements beyond the acceptance criteria. The opposite can also be true, where an entirely exploratory approach might end with a chaos of mind maps and important journeys without automated coverage.

The *Agile Testing Quadrants* are a great example of a balanced testing approach (Crispin2011). There are two dimensions forming the four agile testing quadrants: the focus of the tests ('Business-Facing' or 'Technology-Facing') and the purpose of the tests ('Critique the Product' or 'Support the Team')). Ash created this balanced testing strategy for a client based on the agile testing quadrants:

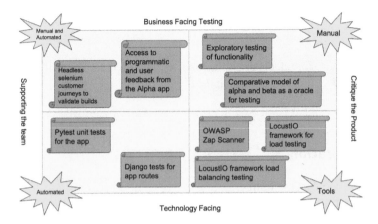

Balanced Testing Approach

With the use of an ephemeral development environment, each activity could be performed, from initial load testing for concurrency problems to running automated acceptance tests locally. To achieve this, we need to set a benchmark for the current development environment and decide what types of testing to enable now, actions needed for that enablement, plus how we might extend it in the future.

We advocate small evolutionary testability changes, rather than large changes which can be hard to justify next to new features and resilience challenges. The following exercise explores how to improve your development environment for testing purposes.

What does the exercise involve?

Follow this facilitator's checklist below to help you run the exercise.

Who do we need?

We suggest involving all members of the team in this exercise. Product and design would be useful here as we are going to discuss which types of testing are essential to enable early feedback on those areas.

 For full details of the roles, see Team Test for Testability exercise in Chapter 1.

Preparation

- Time - One hour
- Type of Space - An open space with a whiteboard
- Attendee Preparation
    - Familiarity with the Agile Test Quadrants model
    - If you have a current version of your development environment running locally, we recommend a session to get it running on the team's computers.
- Physical Materials
    - A whiteboard or flipchart
    - Sticky notes and pens (one color)
- Digital Materials
    - If you have an example of your current development environment, bring a computer running it.

Sample agenda for pre-communication

- Set the Stage (5 mins)
    - Explain the importance of the development environment, its ability to close a feedback loop between developers and testers, as well as between teams and product stakeholders.
- Gather Data (20 mins)
    - Brainstorm the types of testing currently done on your development environment.

- Brainstorm the types of testing required to be enhanced or enabled early in the pipeline.
- Generate Insights (20 mins)
  - Allocate the types of testing into each of the 3 categories in the testing quadrants
  - Determine the first action needed to enable or enhance that capability.
- Agree actions (5 mins)
  - Pick out the actions needed to enable or enhance important testing activities and add to the backlog.

Facilitation

- Draw the Agile Testing Quadrants on a whiteboard or flipchart as per this picture:

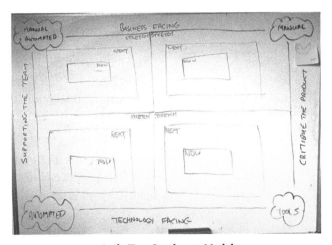

Agile Test Quadrants Model

- There are three categories, explain each:
  - Now - what you must be doing to get early feedback to your context. Only allocate 2 per quadrant here, so as not to set the scope for improvement too wide.

- Next - To get closer to deciding what activities to add to your development environment capabilities, allocate a maximum of 3 per quadrant here.
- Stretch - Basically all the other activities, stored here as reminders to improve in the future while acknowledging that these are not as critical as other activities.

Frequently Asked Questions

- What are types of testing?
  - Use the testing quadrants model as a starting point to trigger ideas for what is currently done and what should be enhanced or added. It is possible you might all agree to enhance what is currently being done which is useful. However, we don't usually find teams performing testing activities in each quadrant - as they should - early in the development life cycle.
- Why do we need to fill in every quadrant?
  - Without balance we can't give useful early feedback to each other and our stakeholders, we will always be giving incomplete feedback on quality, reducing the usefulness of our decisions.
- I'm a product person, why should I care about this?
  - Essentially, we want to give the earliest and most well-rounded information about your products and designs and enable you to give feedback to the team and make meaningful decisions.
- I'm a tester and the application needs to be on a shared test environment before I can meaningfully test it.
  - Testing will still need to be done on test environments throughout your pipeline, this isn't a replacement for that. This enables you, as a tester, to control your environment to a greater extent, and more easily observe the impact of changes in behavior.

Goals

- A comprehensive picture of which types of testing are essential to give faster feedback to all stakeholders.
- Determine actions that will enable the desired testing activities in chronological order of importance.

Outputs

- A model of the Agile Testing Quadrants, with the following layers:
  - Testing activities to enhance or enable now and first steps to doing so.
  - Next stage testing activities to enhance or enable and first steps to doing so.
  - Stretch testing activities to enable.

What do the results mean?

To explain what some of the results might mean, here's an example of a team identifying types of testing activities in the "manual" quadrant focused on business facing tests that critique the product. These represent ways to exercise the product that find subtle customer facing problems that might compromise the value the product can provide:

- Now - More exploratory testing of the HTTP API layer which feeds data to their web application to test early for resilience in the integration between the two.
- Now - The team would like to be able to test feature toggles and combinations of toggles based on what is already enabled in the production environment.
- Next - The ability to test your development environment setup on multiple browsers, devices and operating system combinations on the organization's device farm, Saucelabs.

- Next - Encourage earlier hallway testing by being able to show stakeholders early prototypes on their own devices.
- Stretch - Test changes within and around key customer journeys as early as possible.

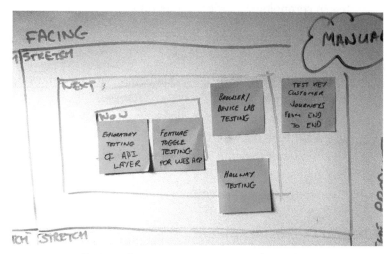

Business facing activities to critique the product

Afterwards, the team identified a few actions to get started with:

- Now - containerize the HTTP API and the feature toggle API as part of the development environment build. Then add the new containers to the already containerized web application and orchestrate them to connect to each other.
- Next - work with the networks team to safely tunnel a development environment to their chosen device cloud and promptly propagate development environment URLs to internal DNS, so anyone can access them.

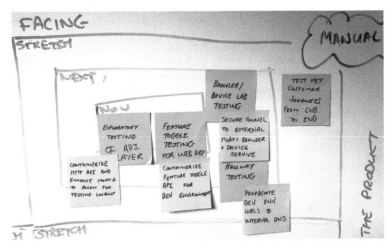

Actions needed to enhance or enable those activities

The team talked about a number of different testing activities and their relative value, committed to improve their development environment and also to begin working more closely with the network team.

We believe the development environment is a crucial collaboration point for all roles within a team and its stakeholders. This approach scales for other environments too, reinforcing that test environments should not exist for their own sake. Instead, they should enable the flow of information from testing to stakeholders, enabling better-informed decision making.

5.4 Summary

In this chapter we have covered the following concepts:

- There are a number of challenges posed by persistent test environments. Sometimes they have been long-standing issues which trigger significant investment.

- The use of ephemeral development environments evolves over time, leveraging them for greater collaboration and flow of work through a team.
- A testable system is one that enables you to perform a varied set of testing activities via a balanced test strategy.
- Creating the conditions for a varied testing strategy that can be executed early, on a development environment, closes feedback loops that previously needed subsequent deployments to later environments to realize.

Test environments have a significant impact on your testability, they can be a real strength or a massive source of pain. Early use of ephemeral environments (close to initial development) over large amounts of testing on persistent environments (much later in the lifecycle) can provide real benefits to team cohesion and flow.

6. Use production data to enhance your testing strategy

Key points

- Static models of risk in production environments **perpetuate assumptions** about the system. This leads to customer-impacting problems and outages.
- A test strategy needs to **evolve as conditions change** in the production environment. We must apply new techniques, amplify current techniques and deprecate ineffective techniques.
- Having an effective **deploy and release strategy** helps control our risk exposure. Testing alone cannot mitigate those risks.

This chapter covers techniques for improving the testability of software through approaches to testing in production (live).

6.1 Static test strategies cause problems in dynamic production environments

Conditions in your production (live) environment change as your product changes, but the way teams test often remains stagnant, unfortunately. This includes the techniques that are in use and the depth to which we apply them. We should regularly ask whether current testing is still effective at all in these new conditions. For us, production is 'where it matters', informing why, what, where and how we test. Value provided by testing has a direct relationship with contextual information from production.

There are many reasons a test strategy needs to change over time. For example, when a company transitions from a start-up-style system to a more established one, customers come to depend on both the stability of the software and the function it provides. In these circumstances, the testing we need to perform changes but there is a danger that the test strategy remains the same. Within larger organizations, growth of a new software feature can change conditions in production. In large organizations changing strategy can be a long process requiring significant advocacy. This leaves a long window for customer problems to occur, as the test strategy does not mitigate the current risks.

Having a static test strategy while conditions in production change leaves teams exposed to an increasing level of risk; even worse, the exposure to risk grows over time. This can manifest itself in the following ways:

- Who does testing - participation in testing across disciplines can decrease. Only a few domain experts hold the knowledge to test.
- How we do testing - the testing activities are ineffective in

context. For example, relying on scripted manual testing in a complex domain.

- What we test - the coverage of testing does not match what the business values.
- What environments we test with - infrastructure changes that happen in production are not reflected elsewhere.
- The data we test with - we use unrepresentative data in earlier environments, either in terms of volume or complexity.
- The tools we use - the ability to inject load for performance and capacity testing are no longer sufficient for example.
- The external environment changes - major new pieces of legislation are not factored in testing as they impact production the most, which is where we test the least.

Production is an 'oracle' or truth-teller for determining how effective our test strategy is; it allows us to evaluate if the risks we identified and mitigated were appropriate (Kaner2012). Better understanding of the information that your production environment emits benefits your testability because your team's knowledge of the environment, application, their users and experience with the product increases. Without this knowledge, the following problems will grow in impact:

- **Persistent Assumptions**: When a test strategy is not informed by production, assumptions persist. Teams assume that users behave in certain ways when operating the application. They believe users follow manuals or instructions which then skews later testing. There may also be assumptions around infrastructure capabilities and behavior of software dependencies.
- **Untested Claims**: Teams and stakeholders make claims about the software they have built. For example, untested performance or capacity limits can be common, resulting in outages. It is difficult to test these claims without a feedback loop from production as they can be true or out of date.

- **Walking the Same Path**: If a test strategy does not change, you revisit the same information over and over. This can lead to problematic areas of the application left unexposed. These areas might have frustrated users inhabiting them which in turn can damage the reputation of the team and the business itself. This all leads to a lack of trust from stakeholders. Our biases often dictate that we build and test features for people like us; by getting data from production, we can determine if our biases are having a detrimental effect on our users.
- **Missing Stakeholders**: If a test strategy does not change, it excludes emergent users. This might be a new team which performs a function in the application or a set of customers from a new territory.
- **Opportunity Cost**: All testing has an opportunity cost. Performing one testing technique means not performing another type. It is the same for test coverage too. All these choices have a cost and are best informed by your production environment.

Identifying your production conditions is the first step to adding a feedback loop. To do this, we need a collaborative method to gather data sources from stakeholders.

6.2 Exercise: employ data from production to keep your test strategy relevant

Great testability is not only how observable, controllable and decomposable your system is. It is also how well the team understands system usage by customers. This understanding helps to target testing to what matters. In this section, we explore how to update a test strategy based on data gathered from the production environment.

Getting everybody together to decide on data sources

Selecting what data sources to use is an important first step in a dynamic, evolving testing strategy. All aspects of test strategy evolve, including the testing activities used and the data that informs them. We recommend that you repeat this exercise every 3 months or so particularly for a long running team or after a major architectural or product shift.

The following types of production data sources are significant influences on test strategy:

Type	Explanation
Performance Trends	Breaching thresholds for load times and responsiveness.
Usage	Sudden changes in demand, a new feature or an uplift in usage of an existing feature.
Access	Method of access to a system may change over time. New mobile devices, authentication libraries or load balancing changes mean access changes often.
Incidents	Number and severity of incidents and outages change our test strategy. This includes where we test, including areas of historical neglect and low coverage.
Customer Support	Number and nature of customer support contacts is a very useful metric. Customer experiences with certain areas or features of your application point to potential confusion or incomplete customer journeys. This could suggest a coverage discrepancy.

Type	Explanation
Deployments	Bringing information about the 'Four Key Metrics' (Riley2015). If deployment frequency is high then team cohesiveness is likely high, which is a positive factor for testing. Alternatively, if lead time is long, change failures are high and time to recover from failure is long then testing might be part of the problem.

How does the exercise work?

Visualization is key to this exercise. We recommend preparing two artefacts for the discussion.

The first is a visualization of your current test strategy. We recommend using the Agile Testing Quadrants for this purpose (as mentioned in Chapter 4). An example is below:

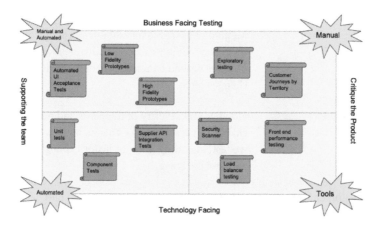

Agile Testing Quadrants

Second, draw up a list of potential data sources available to you, those that give you useful data about conditions within your production environment. These might include the following as a guide to get you started:

- User Centric: you need to be aware of how different types of users interact with the product and what means they use to do so. A business-to-business (B2B) product might have a small number of high-usage customers, whereas an ecommerce website has many customers who individually have a lower impact.
- Architectural Change: Our system architecture is continuously changing as we develop new features and our capacity needs change. Load balancing and innovations such as autoscaling make this dynamic (AWSautoscaling).
- Telemetry: Systems generate events based on human (customer metrics) and machine (diagnostics such as memory and processor usage) actors within the system. This could come from logging, monitoring and alerting for example. These are all valuable sources of feedback for a test strategy.
- Dependencies: The performance and reliability of dependencies constrain our testing. Quality is variable and not within your control with external suppliers. Information such as the uptime of your dependencies or incidents involving their services would be useful here.

Who do we need?

Involve architects, developers, testers, operators and analysts in the exercise. This could be wider for later sessions with product stakeholders who bring information about future features and growth.

 For full details of the roles, see Team Test for Testability exercise in Chapter 1.

Preparation

- Time - 90 minutes

- Type of space - An open space with a large whiteboard
- Attendee preparation
    - Familiarize with the current test strategy
    - Create a list of potential data sources
- Physical materials
    - A whiteboard or flipchart.
    - Sticky notes and pens.

Sample agenda for pre-communication

- Set the stage (10 mins)
    - Check understanding of the current test strategy.
    - Give an example of a data source and a threshold to set the scene.
- Map Data Sources (30 mins)
    - Collate and deduplicate data sources.
    - Classify data sources into the following suggested categories: **Telemetry, Topology, User-Centric, Dependencies**
    - Map data sources to visual representation of test strategy
- Create or Update Thresholds (30 mins)
    - Identify critical data sources based on impact on the team, product and your stakeholders.
    - Add thresholds to each of the data sources you identify.
- Recommended Strategy Changes (20 mins)
    - Check for any exceeded thresholds.
    - Create actions to add any new test strategy changes required.
    - Create actions to remove or reduce any test strategy elements with lesser value.

Facilitation

During the first run-through, pick out a limited number of data sources from a broad range of categories. This will grow over time, allow it to emerge. If there are many changes to the test strategy required, use a dot voting strategy to narrow them down.

When the session is repeated in the future, you will need to:

- Add new data sources to the test strategy
- Add thresholds for change to those
- Check existing thresholds for change

Frequently Asked Questions

- Can a data source or threshold be used for many elements of a test strategy?
  - Yes, and this is very likely to happen. Many of the elements in your test strategy will have the same dependencies, such as common libraries and third-party services.
- What is a good example of a data source with a threshold?
  - We recommend you use browser coverage as a data source (such as Google Analytics Browser Usage). For the threshold, if a browser is above 5% usage, add to the automated acceptance test device list.
- What happens if we need to reintroduce a reprioritized element of our test strategy?
  - We suggest keeping an archive of changes. Store the test strategy in version control or in a wiki format. Review changes before each update.

Goals

- Keep the team test strategy up to date with the latest information from production.
- Increase testing on what matters most.
- Reduce focus on elements of the test strategy with declining value.

Outputs

The outputs from the test strategy update exercise should be an updated test strategy with:

- Data sources and thresholds added to identified test strategy elements.
- Actions created to expand testing where needed.
- Actions created to reduce testing where the value is lesser.

What do the results mean?

After adding data sources and thresholds you should have visualization of your test strategy that looks something like this:

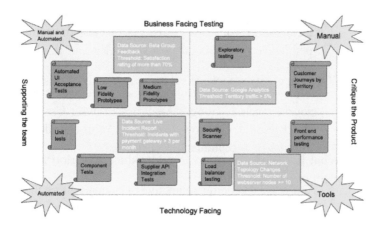

Test Strategy with Data Sources

For each of the testing techniques in each quadrant, the team has added a data source, with a threshold. If we pick out one from each quadrant, we have:

- Business-facing testing that supports the team:
  - Element of Test Strategy: Prototyping

- – Data Source: Beta users feature feedback survey
- – Threshold: At least a 70% satisfaction score for new features.
- Business-facing testing that critiques the product:
    - – Element of Test Strategy: Customer Journeys
    - – Data Source: Google analytics traffic from each active territory.
    - – Threshold: New territories reach 5%. Established territories drop below 5% of total traffic.
- Technical testing that supports the team:
    - – Element of Test Strategy: Integration Testing
    - – Data Source: Live incident report from integrations with payments gateway provider.
    - – Threshold: More than 3 customer impacting incidents per calendar month.
- Technical testing that critiques the product:
    - – Element of Test Strategy: Capacity
    - – Data Source: Network topology documentation for nodes per service. Those currently in live load balancing pool
    - – Threshold: Increase to 10 or above for key services.

These are key metrics for the team and how and what they test. Determining what changes to make based on these thresholds is important. This particular team decided upon the following:

- Prototyping: The satisfaction rating was only at 50%. This suggests that medium fidelity prototypes had not reached enough internal users (Beaubien2019). The team wanted to increase the number of internal users to gather more feedback. Before committing to build.
- Customer Journeys: After reviewing Google Analytics traffic many territories had changed. The team decided to decrease the number of journeys in the FR territory that had declined below 5%. The DE territory needed increased coverage as it was growing.

- Integration Testing: The team found 5 customer impacting problems with their payment gateway. The provider's sandbox environment was problematic too so testing early was challenging. This caused many frustrating build failures. The team agreed to setup a regular call with the provider's support team. The aim is to get a stable environment to test against earlier in the development cycle.
- Capacity Testing: The team hadn't considered this before. Now the application had scaled above ten web servers and was growing fast, it was time to add it to the strategy.

The final visualisation of the strategy looked like this. Giving the team a reference point for future iterations of the exercise:

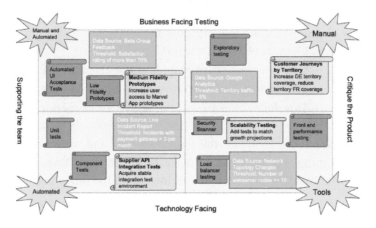

Team Test Strategy with Actions

Essentially, the team built a set of health checks to measure their test strategy against. This exercise is a powerful tool for checking that you are testing what matters, when it matters, with tools and techniques suited to your context.

6.3 Use deploy and release patterns for feedback on high risk changes

A test strategy updated with data from production is important. However, testing alone is not enough to mitigate certain risks, particularly in distributed systems at scale. We have seen teams (and testers) struggle to test profound system changes. We will describe two examples of this for dependency upgrades and technology migrations.

With these types of changes it is critical to control your risk exposure, making gradual changes and gathering feedback based on criteria for success. We can do this by careful intervention using techniques like exploratory testing behind a feature toggle or using chaos engineering (Boyter2016). Passive feedback is possible too, using analytics to see what journeys customers are on and how successful they are.

We can control this risk exposure by using the pattern of separating 'deploy' and 'release', which is key to enhancing our testability. We explore the value of separating deploy and release in the following sections.

Enhance testability through separating deployment and release

As we've seen, testing alone is insufficient to control our risk exposure in production. We also need to decouple the technical decision to deploy and the business decision to release (Humble2010). Between deployment and release, there is an opportunity to gather feedback. This could be from testing or other activities to gather information regarding the risk (or reward) of the change.

When you deploy an update, you install it on your production environment but don't yet expose it to live traffic. To release means to direct some live traffic to the newly-installed software version. Or put another way, releasing moves traffic to the new version (Gillespie2017).

Deploy and release patterns and their testability benefit

There are many deploy and release patterns that your organization might already use. Or it might be that your organization is considering how to put this in place. It's vital to know testability benefits, so you can advocate for the approach that offers the most value.

Pattern	Description	Testability Benefit
Dogfooding (DevIQ2014)	Using the products you build, in the same manner as your customers. Often providing an early internal version of your product for feedback.	Fosters greater empathy with your customers and those who keep the product operational. Wider testing of early versions provides a diverse view before deploy and release.
A/B testing (Optimizely)	Comparing two versions of a feature against each other to see which meets a specified goal.	Shifting the culture from "we believe" to "we measure." Fewer claims and assumptions about the product are present in this culture. The data used to guide testing results in a better targeted strategy, improving flow.

Pattern	Description	Testability Benefit
Beta testing (Wikipedia_Beta)	A version of the product where you invite a private set of customers (closed). Or open the new version up to interested parties.	Both open and closed betas have the testability benefit of providing a wider view of quality. A closed beta gives you the control to invite a particular customer set. Where an open beta gives you access to randomness of usage. This is challenging to inject into testing for many teams.
Blue green deploy (Humble2010)	This approach ensures you have two identical environments. With the ability to switch traffic between them, enabling you to deploy and test on the offline version, plus rollback if you need to after switching over.	This enables teams to experiment within their production environments with new features. This enables testing such as resilience or performance on production with less impact.

Pattern	Description	Testability Benefit
Canary builds (Santo2014)	Use this technique to deploy new software versions to a subset of your infrastructure. After testing, you can then route selected traffic to it to get further feedback.	This gradual release technique gives feedback on performance or capacity sensitive changes. Coupled with an effective test strategy, this can help confirm software dependency upgrades.
Feature flags (Hodgson2017)	This means you can ship many code paths within one deployable. This code path is then exposed to a targeted set of customers.	This is powerful for exploratory testing post-deployment and pre-release. This does come at a complexity cost if not managed well, including nesting feature flags within each other. Requires discipline to remove obsolete code paths to prevent complexity.

Pattern	Description	Testability Benefit
Verify Branch By Abstraction (Fowler2014)	Build an abstraction layer between our application and an entity you want to replace. Over time the entity you want to replace uses this abstraction for all interactions with our application. You then integrate the new entity with the abstraction layer.	An abstraction layer helps to isolate issues. It provides a mockable interface for performance and capacity testing. It can help to replace dependencies that constrain your own testing.

To understand these patterns better, and the testability benefits they provide, we will use two case studies. These show the use of the Verify Branch By Abstraction pattern and canary releasing for high risk changes.

Using the Verify Branch By Abstraction pattern to migrate to a new persistence layer

The *Verify Branch by Abstraction* pattern is effective at enabling teams to use real data at Production volumes. It can turn a challenging, risky change into a routine, stress free experience.

Rob experienced this at a previous client:

A few years ago, I was working at an early-stage software-as-a-service start-up. We were beginning to get traction in the market and were attracting some of the biggest online retailers in the world as customers. This success brought us new challenges – the volume of data that our platform had to handle increased by a factor of ten. Because our service was embedded in a customer's checkout flow, our platform needed to be available, responsive and scalable to customer demand. With this increasing load, we began to experience problems with our platform's persistence layer: it was unable to write and read data [as quickly?] as we needed. Unfortunately, Black Friday, the busiest shopping day of the year, was only a few months away and our platform was at breaking point. We decided that the only way our platform and company could survive Black Friday was to replace our entire persistence layer. This represented a huge challenge with many risks. We had to ensure that:

- *Historical data will still available to both our business and our customers.*
- *The platform was available 24/7 during the transition.*
- *The data was present and correct.*
- *All the existing services that relied on the data still worked.*

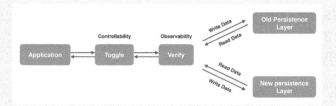

Verify Branch By Abstraction pattern

We couldn't currently replicate production data in our test environment – and the volume, variety and complexity of the data already in our test environment was not enough. We decided to use a software design pattern called Verify Branch By Abstraction. This enabled us to test our new persistence layer [safely?] with real production data and allowed us to control our risk exposure while observing system behaviors.

To reduce our exposure to risk, we added feature flags to control data flow, which enabled us to read and write data from either the old or new persistence layer.

To test data consistency between the new and old persistence layers, we added some instrumentation to the code: with dual-read enabled, we could compare the data read from each, and if it was different, we threw a silent exception for investigation.

We managed to get through Black Friday without any serious outages. By taking a strategic approach to a risky situation, we protected the customer and the teams who supported the system.

Using canary releases to update critical software dependencies

The Canary deploy and release pattern is useful for updating software dependencies or libraries, especially when you are establishing the habit of updating them.

Ash experienced this at a previous client:

We had been busy building new features for a long time. So busy that we had neglected to update the version of PHP that we used for our mobile website. We knew these updates should be an incremental improvement activity. Performance concerns made it a must with large marketing campaigns on the horizon. In fact, out of date dependencies is a great indicator of a lack of testability. After all, we fear these upgrades because of the lack of tests to provide confidence.

Moving two major versions of your core programming language is quite a challenge. Yet that dependency had its own dependencies. MongoDB was our data persistence layer of choice. The version of the Mongo driver we used would not work with the new version of PHP. Lesson number two: dependencies have their own dependencies. The dependencies are always at least a level deeper than you think.

A tester in the team responsible was nervous about what was about to happen. As sometimes happens with hard to test, risky problems the tester was in a difficult position. They had to come up with a test approach for this. But testing alone is not enough to mitigate these risks. We came up with a three-point plan:

Change the areas of the application code in small increments, adding unit tests where code coverage was low (some areas had not changed for a while so there was testing debt present) Deploy to pre-production once enough increment [iterations?] had been completed to the team's "Definition of Done" Deploy the upgraded code and dependencies to ten ring-fenced web servers in production. We would add extra logging and monitoring for performance and resource usage to give us feedback. The web servers could then be introduced to production traffic at sensible times and volumes, while being monitored by the team.

As a result of this approach, we got some useful feedback. Load testing exposed a major problem with how the updated MongoDB driver managed user defined compound indexes (MongoDB). Exposed to enough traffic, this would have caused a live outage. Running the code in production with limited traffic exposed some problems with file input/output. There is more variety in the production traffic, which exposed many smaller issues.

We should be mindful of where testing is not enough to meet our needs. A core aspect of testability is knowing where testing can be most effective, then using other techniques to complement our testing.

6.4 Summary

This chapter covered the value of using data from production to inform and update the test strategy:

- Aligning your test strategy with your production environment prevents assumptions and claims taking precedence over risk calculations.
- Using a balanced test strategy diversifies testing techniques and stakeholders. Using your production environment as a data source gives better value to stakeholders.
- Testing alone is not enough to control risk for the systems that we build and operate. Gathering gradual feedback using deploy and release patterns complements testing.

Distance from production environments introduces weakness into your test strategy because it ignores the potential for feedback from where we get the most insight, both on how we test and what we test for. A production focus increases social cohesion between development teams, customers and operators through a focus on the environment they care most about (production). This is crucial for cultures striving to encourage development teams to build and run software.

7. Use team testing reviews to enable sustainable delivery

Key points

- Without a focus on testability, teams accrue testing debt which slows feedback, diminishes team productivity and increases rework.
- A balanced whole-team test approach allows teams to find problems quickly and cheaply while minimizing testing debt.
- Regularly reflecting on the team's testing experience creates a whole-team awareness of testing debt and an appreciation for the value of testability.
- Teams need to exploit customer-impacting issues to learn about the value of their current testing practices and target areas for improvement.

This chapter covers techniques for improving the testability of software by taking a sustainable approach to software delivery.

7.1 Testing debt affects wellbeing and sustainable delivery

Working in a healthy, sustainable manner is an important ingredient in the productivity and long-term success of your team. Teams will be successful when you create an environment in which they can do their best work\; a safe environment focused on learning and improvement. The teams that we have seen consistently deliver valuable, reliable software while maintaining a sustainable pace have one thing in common, a focus on testability.

When starting out on a project, testability is often low on a team's list of priorities as they focus on building a minimum feature set as quickly as possible. Teams can quickly accumulate substantial amounts of testing debt.

Testing debt occurs when a team deliberately or inadvertently chooses an option that yields benefit in the short term but results in accrued testing cost in terms of time, effort or risk in the longer term (Fowler2009).

When a team builds up substantial testing debt there is an opportunity cost associated, the quality and depth of the testing the team is capable of performing is diminished. As a result, important problems and risks that require thought, time and investigation have a far greater opportunity to remain in the system undetected.

> Rob had an experience that illustrates the benefits of addressing this opportunity cost:
>
> *"Early in my career, I was a tester working on a development team building a 3D visualization product for the safety automation industry. As the product was used in safety-critical environments it had to be tested rigorously and all observed issues had to be reported and investigated. The software product*

would crash regularly while interacting with the user interface, usually as the result of some convoluted pattern of use. Often, we could spend two to three days trying to reproduce a single crash without success, meaning that we were losing weeks of testing and development time every quarter. While I was struggling to reproduce a crash with one of the developers, he suggested that he could add a configuration to capture details of all the interactions with the user interface and write them out to a log file. Once this feature was in place, reproducing these crashes was as easy as reading the log file. The time that I saved investigating these issues allowed me to create a suite of automated user interface tests and to begin testing the product in a variety of ways I had never had time to do before. The overall result was that we were able to do more extensive testing in a shorter period of time."

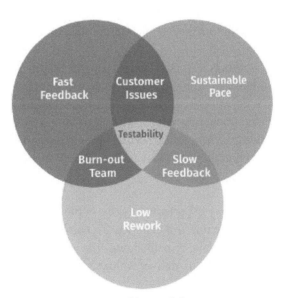

Sustainable Testability

During our careers, we've seen that as the amount of functionality in a system grows, the number of possible states and interactions explode making testing slower and more difficult. At this point, if teams don't prioritize addressing their testing debt they face the following challenges:

1. **Slow Feedback** - Teams are unwilling to compromise on test coverage, deciding that extensive testing must be completed. For example, a team may need to perform extensive manual testing as they didn't maintain a reliable automated regression testing suite. As a result, each subsequent change takes longer to test. Quickly, a team can find that development and testing are out of sync with testing becoming an afterthought. The feedback cycle slows and the time to get a change into production grows with every iteration.

2. **Unsustainable Pace** - Teams react to the growing testing workload by pushing themselves to work outside of normal working hours. This behavior, if maintained, results in significantly diminished team morale and productivity levels (Robinson2005).

3. **High Rework** - Teams scale back the depth and quality of their testing allowing them to release on time, however, this tactic often results in an increase in the number of customer-impacting issues. Managing, investigating, fixing and releasing these issues becomes a huge burden on the team, meaning that they have less and less time to spend on building new features.

Although a team may experience any one of the challenges outlined above, more often teams will experience a combination. Over time, unless this testing debt is paid down, the impact on the teams wellbeing and ability to deliver valuable software can become crippling.

Occasionally, we may choose to accrue testing debt to meet an immediate tactical business goal. However, in order to avoid a situation where this debt becomes a significant obstacle in our team's ability to remain productive, we need to continually identify, quantify and remediate our testing debt.

Rob experienced a situation like this:

"Recently, a key customer issued an ultimatum stating that they would not renew their contract unless we added support for a specific piece of functionality before their annual renewal date two weeks later. As a team, we knew that it would be very difficult to meet the deadline, so we agreed to deliver the functionality without instrumentation (Wikipedia_instrumentation), automated acceptance tests or a monitoring dashboard. As soon as we released the required functionality, we began work on addressing the testing debt we had accrued."

7.2 Adopt a whole-team approach to minimizing testing debt

To prevent testing debt mounting to unsustainable levels we need to establish a whole-team test approach. Teams need to establish a way of working that brings all team members together to identify and mitigate risk, ensuring that testing debt is tackled.

In our experience teams can maintain an optimal development pace indefinitely when testing is a whole-team responsibility and delivered as part of the team's "Definition of Done" (AgileAlliance). The whole team works together to ensure the appropriate blend of automated unit, integration and acceptance tests have been written and the applicable risks have been adequately explored before starting new work.

Relying solely on automated testing, teams often miss issues related to usability, error handling or unanticipated scenarios. If teams rely solely on exploratory testing, they struggle to achieve adequate levels of test coverage across the product, exposing the team to risks. We have learned that to achieve the best return from a team's testing efforts, the team needs to adopt an approach that combines exploratory testing focused on unearthing unexpected risks with automated testing focused on confirming expected behaviour.

Teams need a comprehensive suite of automated regression tests to support refactoring, existing feature evolution and unexpected change detection. Its purpose is not to test what is being built today but to help the team to build new functionality safely tomorrow. This suite of regression tests effectively becomes the team's Andon Cord, alerting the team as quickly as possible to unexpected changes in behaviour (Andon). When the team's automated test suite is owned by the whole team, is delivered as part of development and provides fast, reliable feedback it becomes the foundation that supports exploratory testing.

In this case study, Paul Healy, a Development Team Lead at Poppulo, explains how his team manages testing debt with a combination of automated and exploratory testing:

"Like many teams, we work on a complex software system. We have found that promoting exploration as an activity carried out throughout the software development life cycle mitigates some of the inherent risk in building such a system.

Before we begin development we use exploration to gain knowledge and flesh out stories in our backlog. We try to discover the interactions that exist between modules and components. This coupling isn't always obvious or even intentional. Discovering these interactions at this early stage allows us to scope our stories more effectively and define comprehensive acceptance criteria. This has reduced the number of unanticipated problems that we encounter during development. We also find this a good time to examine UI flows. If we can define better flows and improve the user experience we build that into our stories. This can reduce the number of core flows we need to automate in our UI tests.

During development, we use both automation and manual testing. Automation at unit and service level is for verification and change detection. It verifies that the code behaves as intended and notifies us if that behaviour changes. Once the story is ready, the developer asks a team member to independently review the automation coverage and confirm the acceptance criteria are met in a dev environment. This session is informed by the code review and conversations with the developer. It can uncover new cases that need to be considered.

Once the code has been built and deployed, the focus becomes exploratory testing. We verify the acceptance criteria in a fully integrated environment but ideally most functional issues will have been caught in the earlier session. As developers, we are often guilty of not using our software as our users do in the real world. This exploration is a good way for developers to become familiar with system functionality and user flows, as well as uncovering unexpected behaviour.

After the stories have been completed the team conducts a mob testing session. This is informal and often quick. Its main purpose is to sanity check the development and decide what should be automated in our end-to-end tests. It gives the team a sense of ownership and a high degree of confidence in our work. We don't expect to find issues at this stage - it's far too late in the life cycle.

Exploration is a valuable activity for learning, defining stories, testing the expected and discover the unexpected. A deep understanding of the system is gained through exploration and knowledge of the codebase. This understanding, coupled with comprehensive automation, means developers can move quickly, confidently and sustainably."

7.3 Exercise: use the 10 P's of Testability to track team testing culture

To establish a whole-team testing approach, we first need to establish a benchmark. The team can then iterate towards a more sustainable pace of delivery. This team testing experience exercise encourages team members to highlight challenges and opportunities. Providing perspective to promote transparent and empathetic ways of working. This includes the impact on delivery and wellbeing. Run this exercise at least once a quarter or more often if required to show changes over time.

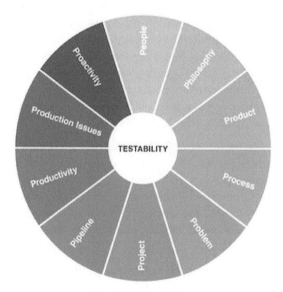

Ten P's of Testability

 Discover more about the 10 P's of Testability in the Appendix.

What does this exercise involve?

Each member of the team will be asked to rate 10 factors that strongly influence their testing experience and to share their rating. The team will then be asked to discuss and agree on a team consensus for each of the 10 factors. Each of the 10 Ps of Testability are listed below.

Title	Description
People	The people in our team possess the mindset, skills and knowledge to do great testing and are aligned in their pursuit of quality.
Philosophy	The philosophy of our team encourages whole-team responsibility for quality and collaboration across team roles, the business and with the customer.
Product	The product is designed to facilitate great automated and exploratory testing at every level.
Process	The process helps the team decompose work into small testable chunks and discourages the accumulation of testing debt.
Problem	The team has a deep understanding of the problem the product solves for their customer and actively identifies and mitigates risk.
Project	The team is provided with the time, resources, space and autonomy to focus and do great testing.
Pipeline	The team's pipeline provides fast, trustworthy and accessible feedback on every change as it moves through each environment towards production.

Title	Description
Productivity	The team considers and applies the appropriate blend of testing to facilitate continuous feedback and unearth important problems as quickly as possible.
Production Issues	The team has very few customer-impacting production issues but when they do occur the team can very quickly detect, debug and remediate the issue.
Proactivity	The team seeks to improve their test approach continuously, learn from their mistakes and experiment with new tools and techniques.

When all team members have voted for a testability factor there is a discussion to agree an overall team rating. Once consensus is agreed by the team the appropriate rating is highlighted. The team can then add an indicator to the trend column to indicate whether the rating is deteriorating, stable or improving since their last session. Any insights gathered during the discussion should be added to the insights column.

Follow this facilitator's checklist below to help you run the exercise.

Who do we need?

We suggest involving all members of the team in this exercise. Product and design would be useful here as we are going to discuss which types of testing are essential to enable early feedback.

 For full details of the roles, see Team Test for Testability exercise in Chapter 1.

Preparation

- Time - One hour (add 15 minutes for your first session)

- Type of Space - An open space with a whiteboard
- Attendee Preparation
    - Distribute the 10 Ps of Testability sheet before the exercise (see Appendix).
- Physical Materials
    - Whiteboard or large desktop.
    - 10 P's of testability cards
    - Sticky notes and markers
    - Results from the previous team testing experience exercise
- Have the whiteboard prepared with the 10 Ps of testability cards in place and the columns drawn before the session.

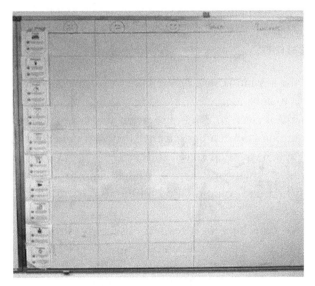

Before 10 P's Exercise

Sample agenda for pre-communication

- Set the Stage (5 mins).

- Explain the goals and FAQ.
  - Introduction to the 10 P's of Testability.
- Gather Data & Generate Insights (35 mins).
  - Rating of each of the 10 testability factors using a simple dot vote in either the unhappy, no opinion or happy column on the board.
  - Ask the team to reach a consensus, plot the trend and capture insights.
- Agree on actions (20 mins).
  - Based on the results, identify which testability factors have the biggest negative impact on the team testing experience.
  - Brainstorm options to address these problems and agree which options you wish to act upon as a team.

Facilitation

- If the session is strictly one hour long, timebox each factor's discussion.
- Actively seek to drive discussion to attain a shared under-standing between participants; focus on understanding why people have rated factors differently.
- Add a downwards arrow, straight line or upwards arrow in the trend column based on the results from the previous session.
- Take some time as a team to discuss the trends associated with each factor. For upwards trends, it's good to explore what the team believes may be contributing to the improvement. For downwards trends, it's good to identify options the team believes may buck the trend.

Frequently Asked Questions

- Does each team member need to have the skills, mindset and knowledge to do great testing?
    - We don't believe that every individual team member has to have the skills, mindset and knowledge to do great testing, but we do believe that the team needs to be able to do good testing in the absence of a testing expert. For this reason, we believe that each team member should continually endeavor to improve their testing skills, mindset and knowledge.
- What happens if the team can't come to a consensus?
    - Usually it's a sign that certain team members have a very different testing experience from others.
    - Those who do more testing will have stronger opinions on the team's testing experience. This is worth exploring further to understand why there is a contrast.
    - If consensus is still hard to reach, opt for the lower rating for a factor because part of the team sees room for improvement.

Goals

- Create a shared understanding of all the factors that influence the teams testing experience as well as the relationship between those factors and the team's productivity.
- Achieve consensus on the factors that inhibit the team's ability to test effectively.
- Generate new insights that may lead to a better overall team testing experience.
- Provide a feedback mechanism to allow the team to evaluate their progress over time.

Outputs

- A snapshot of the state of the team testing experience at a given moment in time.
- Trends for each testability factor that influences the team testing experience over time.
- Insights into how each of the factors inhibits or empowers the team to do great testing.
- Concrete actions that the team can take to improve their team testing experience.

After the exercise you should have a matrix that looks like the following:

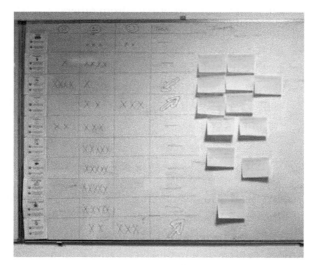

10 P's result

What do the results mean?

The image above shows how each of the team members has voted for each of the 10 testability factors. The team was able to generate the following insights:

- The team testing experience is relatively consistent across the team, meaning that the whole team has a shared understanding of their testing challenges. If members of your team provide very different ratings for the same testability factors this may indicate a lack of communication and collaboration.
- Poor product testability manifested in a lack of observability in some product areas. This lack of observability meant that the team were unaware of data loss used to track customer usage metrics. The exercise indicates this has degraded further since the last time the exercise was run.
- Process and proactivity are good and have improved since the last time. The team has recently added two new lanes to their Kanban board as a result of a team learning review. They have added a Three Amigos lane to discuss the implications of proposed changes and an exploratory testing lane to unearth unexpected problems (Dinwiddie).

Based on this exercise the team decided to focus on improving Product Testability and maintain the good practices that had been established around Process and Proactivity. The team reviewed the insights generated from the Product testability discussion and used the lessons from the Adopt testability mapping to expose common smells of hard-to-test architectures chapter to identify concrete actions. As a result of this exercise the team took a number of actions to add observability to the product.

- They identified their client code, a JavaScript integration for customer websites and mobile applications, as an observability blackspot and subsequently integrated a tool called Sentry which allowed them to see exceptions coming from the client for the first time.
- They improved the structure of their logging, providing machine readability for improved search and filtering capabilities.

- They added instrumentation, monitoring and alerting to each part of the system where it was possible to lose data.
- They created dashboards for their system displaying component status, synthetic test status, error rate and response time.

As a result of their efforts, the team members are now able to assess the health of their system in a fraction of the time it used to take. Issues are now being detected and resolved by the team much more quickly. On the rare occasion that an issue escapes into production the team are often able identify and remediate it before the customer is even aware. These observability improvements have had a significant effect on both morale and productivity of the team.

In many teams, individuals are often unaware of the specific challenges team members face when performing testing activities and the impact on team productivity. This exercise encourages every member to share their perspective, empathize with the challenges faced by their teammates. It reinforces the good work that team has already done and provides guidance as to where the team can improve.

7.4 Exercise: adopt incident reviews to target testability improvement actions

Despite our best efforts, teams will fall victim to a mistake, a customer-impacting issue or a near miss. Testability challenges mean that important problems that impact customers will manifest themselves. Delivery at an unsustainable pace leads to problem deployments, bugs and misunderstandings.

In these situations, the easiest option is to fix the issue and move on soon as possible. When do this we lose an opportunity to learn about our system, how we work and testability challenges.

We propose that teams create a safe space, reflect and learn as much as possible from an incident. We have devised an exercise that helps teams explore what we can learn and what actions to take. We endeavor to learn not only about what went wrong but also amplify what went well.

This learning review exercise focuses on understanding different perspectives of an incident. Drawing inspiration from 'Blameless PostMortems' pioneered at Etsy (Allspaw2012). Identifying decisions, what drove them and how that helps us improve our testability. Within our teams, testing is an information-gathering mechanism for informing our decisions. The quality of our decisions is a product of the quality of our testing.

What does the exercise involve?

The facilitator will start by asking for a high-level overview of the incident before systematically talking through each stage of the incident, asking all attendees to share their experience before seeking to explore actions, obstacles and deep lessons. The stages we suggest using for the review are outlined in the table below. *Note: you may wish to add or remove stages at your own discretion.*

Stage	Description
Detection	How was the incident detected and communicated? Consider who detected the issue, the way it was communicated and the information provided.
Impact	How was the impact of the incident assessed? Consider the impact to customers, the business and the team.
Isolation	How was the root cause of the incident identified? Consider the information available while debugging the incident.

Stage	Description
Fix & Retest	How was the incident fixed and retested? Consider the whole process of getting the fix through the pipeline into Production.
Repair	How was the damage caused by the incident repaired? Consider the repair both from a technical and business relationship perspective.
Impact Minimization	How was the blast radius of the incident minimized? Consider how the extent of the impact was reduced in terms of number of customers impacted, significance of the impact and speed of remediation.
Prevention	How did we try to prevent this kind of incident occurring in Production? Consider the process the team used to identify and mitigate risk.

Who do we need?

We suggest involving the whole team and anyone else directly involved in managing the incident through from detection to remediation. Including operations-focused stakeholders would be really useful here.

 For full details of the roles, see Team Test for Testability exercise in Chapter 1.

Preparation

- Time - One hour thirty minutes (dependent on the complexity & severity of the incident)

- Type of Space - An open space with a whiteboard
- Attendee Preparation
    - Each attendee should review the incident timeline
    - Be prepared to describe the incident from their perspective.
- Physical Materials
    - A whiteboard or a large wall space for post-its.
    - Sticky notes and markers.
- Have the whiteboard prepared before the session if possible.

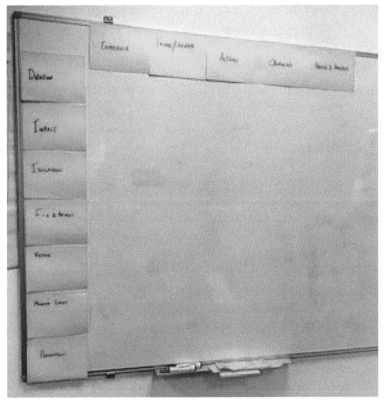

Before Learning Review

Sample agenda for pre-communication

- Set the Stage (5 mins)
  - Explain how the learning review is a safe space where all attendees are encouraged to share their perspective on the incident without fear of blame.
  - Give a general overview of the incident to frame the conversation and ensure everybody knows which incident is being discussed.
- Gather Data & Generate Insights (1 hour 15 mins)
  - Ask the attendees to recount their experiences for each of the stages outlined in the learning review matrix.
  - Discuss learnings, obstacles and actions and capture insights on post-its.
- Plan of action (10 mins)
  - Identify and prioritize actions from the core insights gathered.

Facilitation

- Draw the learning review matrix on the whiteboard listing each of the stages from detection to prevention on the vertical axis and the learnings, obstacles, actions & insights on the horizontal axis. Note: you may choose to add or replace stages on the horizontal axis to meet your specific needs.
- Repeat the following steps for each stage from detection to prevention:
  - Experience - capture all events that took place during that stage of the incident. On-call support logs and messaging apps are good for this.
  - Faster or Cheaper - improvements that would make resolution of that stage faster or cheaper.
  - Actions - capture agreed actions and place them in this column. Use dot voting to narrow the options down.

- Obstacles - record any obstacles that presented themselves, such as access or permissions problems and out of date operational documentation.
  - Process and Principles - changes that apply to your team's ways of working arising from each stage. A change to a team's Definition of Done might be an example.
- Once complete, ask the team to review the action items and lessons around process and principles.
- Prioritize actions, assign an owner and establish a due date.

Frequently Asked Questions

- What if we discover during the session that a team or an individual whose perspective is valuable isn't present?
  - Continue the session and schedule a follow-up session to include the missing party.
- When should we schedule a learning review?
  - A learning review should be scheduled as soon as possible after the problem has been fully remediated. This means that we have a full picture of what happened while it's still fresh in the memory.

Goals

- Gather and share diverse perspectives on the incident to gain an understanding of the decisions made and the circumstances that fueled those decisions.
- Validate the things that went well and identify areas for improvement.
- Generate a prioritized list of concrete actions for the team.
- Uncover important principles by which the team agrees to work.
- Identify process changes required to improve the system of work.

Outputs

- A prioritized list of concrete actions for the team.
- List of new principles the team will adhere to.
- Changes to the team's process.

After the exercise you should have a matrix that looks something like the following:

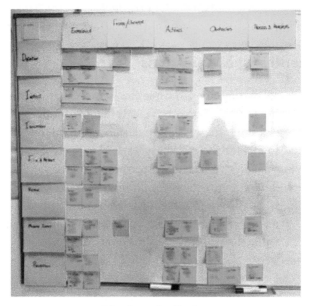

After Learning Review

What do the results mean?

Each one of the post-its in the image above represents a valuable piece of feedback from an attendee. The green post-its in this case represent positive experiences while the pink indicate the negative. It's clear from the exercise that there were many things that the team are doing well and should continue but also a number of opportunities for improvement. In this case, both the prevention and detection stages stand out as areas that need attention as they resulted in the most negative feedback.

One of the most important results of this exercise are insights generated related to process and principles. In this case we have deep lessons at four of the five stages. This particular session was as a result of a minor security issue.

In the detection stage the team noted:

- The team should be able to proactively identify suspicious activity.
- A penetration test that reports no major issues doesn't mean your system is secure.

In the fix and retest stage the team noted:

- The team needs to be able to reproduce the issue before a fix can be pushed to production.

In the minimize impact stage the team noted:

- Isolation and sandboxing of customers' data is a good practice.

In the prevention stage the team noted:

- The team needed clear guidelines on handling security related issues.
- Well-tested third party libraries can be more secure than homegrown implementations.

A timeboxed conversation around each individual stage encouraged the team to focus solely on a very specific set of activities. By limiting the scope to a small, specific set of activities the team were able to explore more deeply. This exercise helped the team to identify concrete actions and lessons, and provided a mechanism for the whole team to learn together.

7.5 Create a board to visualize & prioritize testing debt

Feedback from incidents and outages will give you insight into your testability challenges. We hope they have exposed areas for improvement within your team and system. These improvement areas will likely include testing debt. This could include components without unit tests or dependencies with poor testability (Winter2018).

This debt leads to waste as the team builds and operates their system. Teams need to minimize testing waste while delivering to a known quality. Long release testing cycles are a common example of this type of waste (Smith2015).

The next challenge is how to visualize this debt. By visualizing testing debt on a physical board, teams and stakeholders are conscious of its impact. Teams can use this board to prioritize and justify ongoing investment in testability.

We suggest that your team creates a physical testing debt board where team members can add or update existing testability issues whenever they encounter a problem:

- Keep the testing debt board within sight of the team's Kanban board for development work. Transfer work from the testing debt to the development board when ready.
- Teams can take the findings from their testability exercises and add them to the board.
- Encourage team members to add tickets when they encounter debt that makes testing difficult, slow or unreliable.
- If a ticket already exists for the issue, add a dot so that the team can see how often it manifests itself.
- At the end of an iteration, gather all the team members and review each of the tickets as part of the team's retrospective.

- Rank the tickets' impact on team productivity considering the importance, frequency and effort. Use the 'Three Amigos' method here to reach consensus (Dinwiddie).
- Estimate the investment required to address each of the issues. Use the team's current sizing or estimation process for consistency.
- We recommend following 'The 20% Rule' for debt that the team accumulates when building software (KimHumbleDebois2016). Spend at least this amount of time on your debt, including testing debt.

An example of a board is shown below. This is after a team had completed the team testing experience exercise earlier in the chapter:

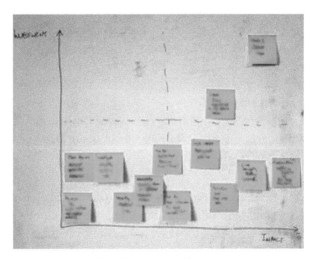

Investment versus Impact

The team can identify which testing debt to address based on maximizing their return on investment. This approach allows the team to make targeted, small improvements in every iteration while still delivering value.

Every team has a certain amount of testing debt but often we have no idea how much testing debt we have accrued. Ongoing focus on testing debt will stimulate team productivity and morale. Providing a visual representation of testing debt opens up the conversation. The opportunity for the whole team to improve areas of frustration follows. The more teams focus on testing debt the less likely they are to introduce it.

7.6 Summary

This chapter covered how:

- Narrow test approaches increase testing debt and customer-impacting incidents. More diversity in testing types performed exposes risks to sustainability.
- Exploring and improving the testing experience fosters empathy and better team cohesion.
- Learning from incidents and outages gives us valuable insights into testability. This exposes assumptions and obstacles in how teams work.
- Testing debt accumulates at the cost of team morale and pace of delivery. Visualization helps us to begin the conversation on how to tackle debt.

Sustainable testability feeds the wider sustainability of all stakeholders involved with the team. These regular ceremonies give the team a voice to talk about the testing they perform, which promotes ownership of testability, which in turn helps stakeholders understand the impact of testability on software delivery. Remember: the aim is reliable software delivery with sustainable testability at its core.

Terminology

- **Acceptance testing**: testing concerned with meeting user needs, requirements and business processes. We generally mean automated in the context of the book.
- **Balanced testing**: performing a range of testing activities that get a broad view of quality. We may also say varied.
- **Codebase**: where the code is kept. We may also say source or version control.
- **Configuration**: referring to configuration files, where software and hardware variables are stored.
- **Controllability**: The ability to control the system to visit each of the system's important states.
- **Cross-functional team**: a team that has programming, analysis, testing and operations skills. We may also say development team.
- **Debugging**: the process of finding and resolving problems within a program or system.
- **Decomposability**: the ability to divide the system into independently-testable components.
- **Dependency**: anything the team depends upon to build and operate its own system. This could be another team, an internal system or a third party service.
- **Design**: usually we are referring to the role of design. Which may include user interface and interaction roles.
- **Development environment**: where developers can program, build and test the application on their local machine. Also known as developing locally.
- **Doubles**: a generic term for replacing a system object or output for testing purposes. We may also say mock or stub.
- **Ephemeral**: intentionally made to last for a short period.

- **Exploratory testing**: simultaneous exploration and learning performed on a system or artefact.
- **External**: something that the team depends upon but is outside of their organization. We may also say 3rd Party.
- **GUI**: Graphical user interface. We may also use UI.
- **'ilities'**: a quality of a system that applies across a set of functional or system requirements.
- **Observability**: the ability to see everything important in the system.
- **Operability**: how well a software system works when operating in production (live).
- **Operational Features**: an alternative coined by Matthew Skelton to non-functional requirements.
- **Production**: the live environment where customers or users interact with the software.
- **Regression**: where software regresses to a previous state. Where a bug is reintroduced that was previously fixed is a common example.
- **Roadmap**: a flexible planning technique to support strategic planning
- **Simplicity**: how easy the system is to understand.
- **Solution architecture**: describes the process of developing, documenting and collaborating with stakeholders over a solution to enable a specific outcome.
- **Synthetic testing**: a method of understanding a user's experience of your application by predicting behavior.
- **Testability**: the degree to which a product or system supports testing in a given context.

References

 Due to limitations of the LaTeX text processor used by Leanpub to generate the book from the manuscript files, some of the URLs in this section may not display correctly. A full list of references is available on the book website.

Visit TestabilityBook.com for details.

Chapter 1 - Testability Priorities

Shilman - David Shilman, Solution Architecture vs Software Architecture, https://dzone.com/articles/solution-architecture-vs-software-architecture

Spolsky2000 - Joel Spolsky, The Joel Test, https://www.joelonsoftware.com/2000/08/09/the-joel-test-12-steps-to-better-code/

Winter2017 - Ash Winter, The Team Test for Testability, https://www.testingisbelieving.blogspot.co.uk/2017/08/the-team-test-for-testability.html

Chapter 2 - Testability and Dependencies

Forte - Tiago Forte, Theory of Constraints 102: The Illusion of Local Optima, https://medium.com/praxis-blog/theory-of-constraints-102-

local-optima-3ca8d348f146

Goodman - Michael Goodman, Systems Thinking: What, Why, When, Where, And How?, https://thesystemsthinker.com/systems-thinking-what-why-when-where-and-how/

Chapter 3 - Testability Mapping

Dinwiddie - George Dinwiddie, The Three Amigos Strategy of Developing User Stories, https://www.agileconnection.com/article/three-amigos-strategy-developing-user-stories

McKee - Lynn McKee, Testing Mnemonics List, Quality Perspectives http://www.qualityperspectives.ca/resources/#mnemonics

Waterhouse - Peter Waterhouse, Monitoring and Observability - What's the Difference and Why Does It Matter?, https://thenewstack.io/monitoring-and-observability-whats-the-difference-and-why-does-it-matter/

Chapter 4 - Architectural Testability

Fowler2014a - Martin Fowler, Circuit Breakers, https://martinfowler.com/bliki/CircuitBreaker.html

Gource - Gource, http://gource.io/

Henderson - Joel Parker Henderson, Architecture Decision Record, https://github.com/joelparkerhenderson/architecture_decision_record

Chapter 5 - Ephemeral Development Environments

Crispin2011 - Lisa Crispin, Using the Agile Testing Quadrants, https://lisacrispin.com/2011/11/08/using-the-agile-testing-quadrants/

Chapter 6 - Use Production to Enhance Test Strategy

AWSautoscaling - AWS Autoscaling, Amazon Web Services, https://aws.amazon.com/autoscaling/

Beaubien2019 - Sean Beaubien, Awesome Guide to Prototyping in User Interface Design, https://careerfoundry.com/en/blog/ui-design/the-value-of-prototyping-in-ui-design/

Boyter2016 - Ben Boyter, What is Chaos Testing / Engineering, https://boyter.org/2016/07/chaos-testing-engineering/

DevIQ2014 - DevIQ, Dogfooding, https://deviq.com/dogfooding/

Fowler2014b - Martin Fowler, Branch by Abstraction, https://martinfowler.com/bliki/BranchByAbstraction.html

Gillespie2017 - Art Gillespie, Deploy != Release, https://blog.turbinelabs.io/deploy-not-equal-release-part-one-4724bc1e726b

Hodgson2017 - Pete Hodgson, Feature Toggles (aka Feature Flags), https://martinfowler.com/articles/feature-toggles.html

Humble2010 - Jez Humble, Patterns for Low Risk Releases, https://continuousdelivery.com/implementing/patterns/

Kaner2012 - Cem Kaner, The Oracle Problem and the Teaching of Software Testing, http://kaner.com/?p=190

MongoDB - MongoDB, Compound Indexes, https://docs.mongodb.com/manual/core/index-compound/

Optimizely - Optimizely, A/B Testing, https://www.optimizely.com/uk/optimization-glossary/ab-testing/

Riley2015 - Chris Riley, Metrics for DevOps, https://devops.com/metrics-devops/

Santo2014 - Danilo Santo, Canary Builds, https://martinfowler.

com/bliki/CanaryRelease.html

Wikipedia_Beta - Wikipedia, Open and Closed Beta, https://en.
wikipedia.org/wiki/Software_release_life_cycle#Open_and_closed_
beta

Chapter 7 - Team Testing Reviews for Sustainable Testability

AgileAlliance - Agile Alliance, Definition of Done, https://www.
agilealliance.org/glossary/definition-of-done/

Allspaw2012 - John Allspaw, Blameless PostMortems and a Just
Culture, https://codeascraft.com/2012/05/22/blameless-postmortems/

Andon - Andon, Toyota Production System guide, https://blog.
toyota.co.uk/andon-toyota-production-system

Fowler2009 - Martin Fowler, Technical Debt Quadrant, https://
martinfowler.com/bliki/TechnicalDebtQuadrant.html

KimHumbleDebois2016 - Gene Kim, Jez Humble, Patrick Debois,
The DevOps Handbook, http://it-books.club/books/the-devops-handbook/
isbn/1942788002

Robinson2005 - Evan Robinson, Why Crunch Modes Doesn't Work:
Six Lessons, https://www.igda.org/page/crunchsixlessons

Sandboxing - Sandboxing Definition, Open-source error tracking
and exception handling Tool, https://techterms.com/definition/sandboxing

Sentry - Sentry, Open-source error tracking and exception handling
Tool, https://sentry.io/welcome/

Smith2015 - Steve Smith, Release Testing Is Risk Management The-
atre, https://www.continuousdeliveryconsulting.com/blog/release-
testing-is-risk-management-theatre/

Wikipedia_instrumentation - Wikipedia, Instrumentation (computer programming), https://en.wikipedia.org/wiki/Instrumentation_(computer_programming)

Winter2018 - Ash Winter, Overcome painful dependencies with improved adjacent testability, http://testingisbelieving.blogspot.com/2018/08/overcome-painful-dependencies-with.html

Appendix - Notes on 10 P's of Testability

1 - People

The people in our team possess the mindset, skill set and knowledge to do great testing and they are aligned in their pursuit of a shared quality goal.

Mindset

Each member of the team feels motivated, fulfilled and is focused on delivering a high-quality product. Team members understand that quality is a whole-team responsibility, appreciate that testing provides critically-valuable feedback, strive to facilitate better testing, shorten the feedback loop and endeavor to prevent defects over finding them.

Skillset

Each member of the team has the skills and experience necessary to perform risk analysis, exploratory testing, write unit, integration and end-to-end tests. The team also has access to a testing specialist with deep testing expertise should their expertise be required.

Knowledge

Each member of the team either has adequate knowledge or has a means of accessing adequate knowledge of the problem domain, technical domain, testing tools and techniques required to do great testing.

Alignment

No one individual on the team is responsible for quality, the team has a shared vision of quality and work together to build quality in, facilitate better testing and to improve the team's way of working.

2 - Philosophy

The philosophy of our team encourages whole-team responsibility for quality while building trusting, collaborative relationships across team roles, the business and with the customer.

Whole-team responsibility for quality

All team members actively identify and mitigate risks, consider testability during architectural discussions, collaborate on testing, prioritize the investigation and resolution of automation failures over new feature work and distil as much learning as possible from customer-impacting issues.

Collaborative relationships

Team members work really closely, making changes to the code to facilitate better testing as well as helping each other complete testing and automation tasks. Each team member talks regularly with people from the wider business and the customer in order to gain a better understanding of the stakeholders' needs.

3 - Product

The product is designed to facilitate great exploratory testing and automation at every level of the product.

Designed to facilitate exploratory testing

Team members can quickly and easily set-up whichever test scenarios they wish to explore and evaluate whether or not the system is behaving as desired.

Designed to facilitate automation

Team members can write fast, simple and reliable automation that is targeted at the appropriate level. The majority of the automation is written at unit and integration level with only a bare minimum written at end-to-end level.

4 - Process

The process helps the team recognize risk, decompose work into small testable chunks, and discourages the accumulation of testing debt while promoting working at a sustainable pace.

Recognize risk

Team members are encouraged to identify risks as early as possible so that they may be mitigated in the most appropriate manner.

Small testable chunks

The team works together to create a shared understanding of what needs to be built and slices the work into small testable chunks with clearly defined acceptance criteria.

Testing debt

Team members work together to ensure all the necessary testing activities are completed and findings are addressed before moving onto the next iteration.

Sustainable pace

The team works together to ensure each chunk of work is adequately tested before moving onto new work. Overtime and out-of-hours work is actively discouraged.

5 - Problem

The team has a deep understanding of the problem the product solves for their customer and actively identifies, analyzes and mitigates risk.

Customer problem

Each team member is constantly improving their understanding of who the customer is, what the customer values, their challenges, needs and goals. This knowledge enables team members to better recognize potential threats to the value of the solution.

Risk

Team members have a deep understanding of their context which allows them to analyze business and technical risk, consider the potential impact of failure and mitigate it with the most appropriate techniques.

6 - Project

The team is provided with the time, resources, space and autonomy to do great testing.

Time

The team is provided with the freedom required to think, prepare and perform all the testing activities deemed necessary to mitigate the risks identified without being put under time pressure or working outside of normal working hours.

Resources

The team has access to the information, test data, tooling, infrastructure, training and skills necessary to achieve their testing goals.

Space

The team is provided with the space to focus on completing their testing tasks without too many distractions and minimal context switching.

Autonomy

The team is given the autonomy to find their own solutions to testing challenges.

7 - Pipeline

The team's deployment pipeline provides fast, trustworthy and accessible feedback on every change as it moves through each environment towards production.

Feedback

The team members are confident that the various forms of automated testing provide comprehensive test coverage, detect functional regressions and provide feedback that's reliable, timely and actionable.

Environment

The team can deploy a change into a production-like environment on demand and can safely perform a range of testing activities including resiliency testing, performance testing, exploratory testing, and so on.

8 - Productivity

The team considers and applies the appropriate blend of testing to facilitate continuous feedback and unearth important problems as quickly as possible.

An appropriate blend of testing

The team works together to identify risk and take a holistic approach to mitigating risk using the appropriate combination of pre-production and production testing. The team uses a blend of targeted unit, integration, end-to-end, exploratory and nonfunctional testing to find problems as quickly as possible. These approaches are supplemented with the appropriate level of logging, monitoring, alerting and observability in production.

Continuous feedback

The team breaks their work down into tiny testable chunks, pairs or mobs on coding, automation and testing tasks and seeks stakeholder feedback as early as possible.

9 - Production Issues

The team has very few customer-impacting issues but when they do occur the team can very quickly recover.

Customer-impacting issues

The team uses an effective test strategy that ensures the majority of issues are either prevented or detected before escaping into production. This means that the team spends very little time firefighting customer-impacting issues.

Recovery

The team has built the system with monitoring and alerting that allows team members to detect production issues before they impact the customer. When issues are detected, adequate logging, observability and reversibility is in place to quickly debug and remediate.

10 - Proactivity

The team proactively seeks to continuously improve their test approach, learn from their mistakes and experiment with new tools and techniques.

Continuously improve

The whole team regularly reflects on how effective their test approach is, discussing activities that are valuable, wasteful or need improvement and takes action where necessary.

Learn from their mistakes

The whole team reviews each costly mistake in an effort to distill as much learning as possible, to identify and address gaps in the team's testing efforts.

Experiment

Each team member is encouraged to learn about testing tools and techniques and is supported in experimenting with new ideas that they believe may benefit the team.

About the authors

Ash Winter

I'm Ash Winter, a consulting tester and conference speaker, working as an independent consultant providing testing, performance engineering, and automation of both build and test. I have been a team member delivering mobile apps and web services for start-ups and a leader of teams and change for testing consultancies and their clients. I spend most of my time helping teams think about testing problems, asking questions and coaching when invited. I am most proud of being a co-organizer for the Leeds Testing Atelier, a free full-day community testing workshop. Its mission is to give those involved in testing a platform to share their stories, particularly those who haven't been heard before.

Rob Meaney

I'm Rob Meaney. I came to work in the software industry soon after finishing college (where I gained a degree in electrical & electronic engineering) and began working as a tester without even knowing what testing was. I learned my trade testing desktop applications for the manufacturing safety automation industry. Soon I got bored of manually checking the same thing over and over to I decided to try automating some of my tests. Since then, I have worked in start-ups, gaming, data storage, medical, fraud detection and communication companies building test and automation frameworks. I've worked as a manual tester, automation architect and test manager. I love testing and read and learn about testing and development continuously.

Index

CPSIA information can be obtained
at www.ICGtesting.com
Printed in the USA
LVHW060808131121
702940LV00013B/13